Services Unbound

EAST ASIA AND PACIFIC DEVELOPMENT STUDIES

EAST ASIA AND PACIFIC DEVELOPMENT STUDIES explore economic developments in one of the most vibrant regions of the world. Each volume in the series presents a blend of rigorous analysis, relevant examples, and lessons from the region's unique experiences. From the intricacies of the services sectors to the factors driving firm productivity, and from the challenges facing health care to the complexities of the green transition, this series will be a resource for scholars, policy makers, and practitioners alike.

TITLES IN THE SERIES

Services Unbound: Digital Technologies and Policy Reform in East Asia and Pacific (2024)

Fixing the Foundation: Teachers and Basic Education in East Asia and Pacific (2023) (World Bank East Asia and Pacific Regional Report)

East Asia and Pacific Development Studies

Services Unbound
Digital Technologies and Policy Reform in East Asia and Pacific

Contents

Out of the Boxes

Figures

Foreword

In today's evolving global economy, services have emerged as a force for innovation, trade, and economic growth. In East Asia and Pacific too, new digital technologies are unleashing the potential of services in a region known for manufacturing-led growth. Over the past decade, services have become key contributors to aggregate labor productivity growth. Services exports have grown faster than goods exports. Also, the growth of foreign direct investment in services has exceeded that in manufacturing by a factor of five in China, Indonesia, Malaysia, the Philippines, and Thailand.

This book explores the profound impact of services on development and the opportunities they present. For example, in the Philippines, the adoption of software and data analytics by firms increased the productivity of firms by 1.5 percent on average over the 2010–19 period. In Vietnam, the reduction in policy barriers such as restrictions on foreign entry and ownership in transport, finance, and business services led to a 2.9 percent annualized increase in value-added per worker in these sectors over the 2008–16 period. The elimination of such barriers also led to a 3.1 percent increase in labor productivity in the manufacturing enterprises that use these services, benefiting small and medium private enterprises most significantly.

The combination of reforms to unleash services and digitalization is not only creating new opportunities, it is also enhancing the capacity of people to take advantage of these opportunities. For example, distance education and telemedicine supported by well-selected, trained, and motivated local staff have led to better learning and health outcomes in the region, although there remains significant inequality in access.

This book emphasizes the need for a balanced approach to liberalization and regulation of services. It calls for the removal of policy restrictions that inhibit competition between service providers, domestic and foreign. It also calls for the establishment of a regulatory framework that addresses market distortions, particularly the problems of privacy, cybersecurity, and monopolies that are emerging in the digital era. Additionally, the book highlights the importance of international cooperation both to facilitate market opening and to address market failures.

As the readers navigate through the pages of this book, they will find evidence of the transformative power of services and the contribution they can make to inclusive and sustainable development. It is my hope that policy makers, researchers, and practitioners alike will find inspiration and insights within these chapters to shape policies to harness the full potential of services for the benefit of economies and societies.

Manuela Ferro
Vice President
East Asia and Pacific, World Bank

Acknowledgments

This report is a product of the Office of the Chief Economist, East Asia and Pacific Region of the World Bank. It is a revised and expanded version of the special focus on "services for development," originally featured in the World Bank East Asia and Pacific Economic Update (October 2023).

The team was led by Alessandro Barattieri and Aaditya Mattoo. The core team members were Omar Arias, Sebastian Eckardt, Daisuke Fukuzawa, Ergys Islamaj, Duong Trung Le, Jonathan Timmis, and Anna Twum. Other members of the team were Yu Cao, Caroline Gerd G De Roover, Young Eun Kim, Veronica Sonia Montalva Talledo, Narya Ou, and Cecile Wodon.

The report benefited from significant contributions by Cristian Aedo, Diego Ambasz, Nguyet Thi Anh Tran, Nina Arnhold, Andrea Barone, Tania Priscilla Begazo Gomez, Rong Chen, Ji Eun Choi, Xavier Cirera, Alba Suris Coll Vinent, Tatiana Didier, Michael Drabble, Dominik Englert, Kebede Feda, Jaime Frias, Federico Ganz, Jun Ge, Serene Ho, Hans Ishrat, Magdeleine Joscelyn, Sabreen Khashan, Saloni Khurana, Tatiana Kleineberg, Csilla Lakatos, Blair Edward Lapres, Dara Lengkong, Neni Lestari, Mariem Malouche, Jessie McComb, Christopher Miller, Graciela Miralles Murciego, Harish Natarajan, Linh Bao Nguyen, Flavio Porta, Lauri Pynnonen, Luis Andres Razon Abad, Isabelle Rojon, Rico Salgmann, Arpita Sarkar, Katherine Anne Stapleton, Wushuang Shen, Lars M. Sondergaard, Francesco Strobbe, Louise Twining-Ward, Ralph Van Doorn, Gonzalo J. Varela, Jessica Rose Wilson, Noah Yarrow, and Juni Tingting Zhu.

Manuela V. Ferro provided valuable guidance and helpful comments. We are grateful for stimulating discussions and comments from Michael Corlett, Tim L. De Vaan, Ndiame Diop, Nicolo Fraccaroli, David Gould, Naoto Kanehira, Ketut Ariadi Kusuma, Rafaela Martinho Henriques, Lars Moller, Lalita M. Moorty, Rinku Murgai, Zafer Mustafaoglu, Owen Nie, Vidaovanh Phounvixay, Cecile Thioro Niang, Agustin Samano Penaloza, Kersten Kevin Stamm, Viet Quoc Trieu, Casey Turgesson, Carolyn Turk, Mahesh Uttanchamdani, Ekaterine T. Vashakmadze, Marius Vismantas, Mara Warwick, Shichao Zhou, the staff of the EAP region who participated in the review meetings on August 3 and August 30, 2023, and the

EAP Regional Management Team meeting on August 31, 2023. We are grateful to Juan Marchetti (World Trade Organization) for serving as external reviewer. We also appreciate the support provided by Geetanjali Chopra, Mariana Lucia De Lama Odria, Mark Felsenthal, and Jeffrey Bryon Sparshott. Mary Fisk and Patricia Katayama provided excellent advice and guidance on the publication process. Mayya Revzina provided assistance with the copyrights. Guillaume Musel designed the cover and the graphics. Nora Mara edited the report. The team also thanks others who have helped prepare this report and apologizes to any who may have been overlooked inadvertently in these acknowledgments.

This report features spotlights on specific country experiences, entitled "Out of the Box." For contributions to these sections, we would like to thank Lamiaa Bennis, Kislay Chandra, Tiger Fang, Dominic Jacques, Carlo Lim, Patricia Lim, Hoan Nguyen Khai, Trismawan Sanjaya, and Shiju Varghese for the insightful discussions and for agreeing to be featured in the report.

Abbreviations

AI	artificial intelligence
AIML	artificial intelligence and machine learning
ASEAN	Association of Southeast Asian Nations
B2B	business-to-business
B2C	business-to-consumer
BPO	business process outsourcing
CBPR	Cross-Border Privacy Rules
CPTPP	Comprehensive and Progressive Agreement for Trans-Pacific Partnership
EAP	East Asia and Pacific
edtech	education technology
EMDE	emerging markets and developing economies
EU	European Union
FDI	foreign direct investment
fintech	financial technology
FSOL	Financial Sector Omnibus Law (Indonesia)
GDP	gross domestic product
GHG	greenhouse gas
GNI	gross national income
HIC	high-income country
HR	human resources
ICT	information and communication technology
ILO	International Labour Organization
IMO	International Maritime Organization
IoT	internet of things
ISIC	International Standard Industrial Classification of All Economic Activities
IT	information technology
LLM	large language model
LMIC	lower-middle income country

Mbps	median download speed
MSME	micro, small, and medium enterprise
OECD	Organisation for Economic Co-operation and Development
PMS	property management system
PP	purchasing power parity
PSA	Public Service Act (Philippines)
PSTRI	Preferential Services Trade Restrictiveness Index
RCEP	Regional Comprehensive Economic Partnership
Rp	Indonesian rupiah
SaaS	Software as a Service
SME	small and medium enterprise
SOE	state-owned enterprise
STEM	science, technology, engineering, and mathematics
STRI	Services Trade Restrictiveness Index
TCS	Tata Consultancy Services
TFP	total factor productivity
TVET	technical and vocational education and training
UMIC	upper-middle-income country
VC	venture capital
WTO	World Trade Organization

Overview

Background

Services are often seen as slow to innovate, tough to trade, and hard to liberate from regulatory restrictions. A combination of technological change and policy reform, however, is transforming services into the most dynamic part of many economies. In East Asia and Pacific (EAP), a region associated with manufacturing-led growth, services are already driving development.

Consider seven facts about services. First, over the past decade, the share of services in economic activity has increased significantly: in output, from 44 percent to 53 percent in China and from 44 percent to 48 in the rest of the EAP economies; in employment, from 35 percent to 47 percent in China and from 42 percent to 49 percent in the rest of EAP.

Second, in the most recent years, the contribution of the services sector to aggregate labor productivity growth has been higher than the contribution of manufacturing in all major economies in the region (figure O.1). Furthermore, productivity in certain services—such as business services, finance, and communications—is higher than in manufacturing, although most services jobs in the region are still in low-skill and low-productivity services like traditional retail and transportation.

Third, services currently represent the most dynamic elements of international trade and foreign direct investment in the EAP region. In all countries but Viet Nam, the growth of services exports exceeded the growth of manufacturing exports in the period 2010–19, especially in digitally delivered services (figure O.2). In most countries, in the period 2012–19, the growth of foreign direct investment in services exceeded its growth in manufacturing by a factor of five.

Fourth, the importance of services will increase further as rising incomes and aging populations across the region shift consumer demand toward services.

FIGURE O.1   Services overall are key contributors to aggregate productivity growth

Contribution to aggregate productivity growth, by sector, selected EAP countries, 1991–96, 2000–06, and 2011–18

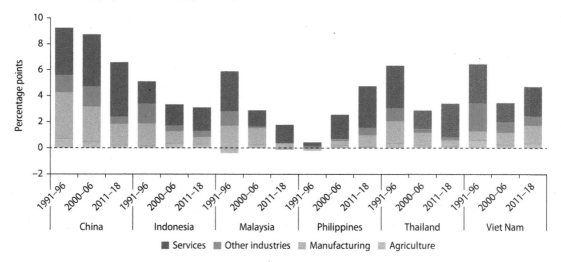

Source: Groningen Growth and Development Centre and United Nations, GGDC/UNU-WIDER Economic Transformation Database (Kruse et. al. 2023).
Note: Figure shows sectoral contribution to labor productivity growth, annual average during three different periods. Other industries include mining, utilities, and construction. EAP = East Asia and Pacific.

FIGURE O.2   Services outpaced manufacturing in terms of growth of trade and FDI in recent years

Growth of exports and FDI inflow, manufacturing vs. services, selected EAP countries

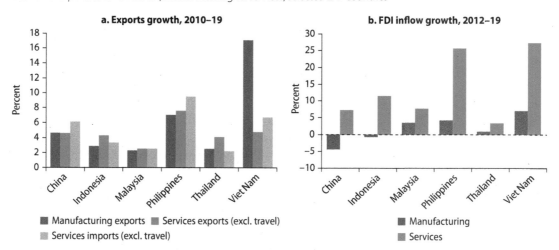

Sources: Haver Analytics; World Bank, World Development Indicators.
Note: Panel a reports the annual growth in the export and import of services for several EAP countries. For this analysis, the services sector excludes travel. The growth rate of manufacturing exports is included for reference. Panel b shows average annual growth of three-year moving-average FDI values, during 2012–19, except for Viet Nam (2014–19) and Indonesia and the Philippines (2013–19). EAP = East Asia and Pacific; FDI = foreign direct investment.

Fifth, because services tend to employ more skilled workers than manufacturing or agriculture, the growing share of services sectors will increase the relative demand for skilled workers. Sixth, services tend to have higher female-to-male employment ratios, and those ratios grow faster in services than in manufacturing as the level of development increases (figure O.3). Last, services, except for transportation, have significantly lower greenhouse gas emissions than industry and agriculture for every unit of output generated. Therefore, structural transformation toward services would contribute to the region's shift to lower carbon growth.

Next, consider the central argument and organizing framework of this report (figure O.4). Changes in technology and services policy are influencing the evolution of services and their contribution to development. Even the uneven diffusion of digital technologies and the limited reform of policies restricting competition have led to structural change across and within sectors. The result is higher productivity in services sectors and the manufacturing sectors that use those services,

FIGURE O.3 **The services sector has higher female employment shares than manufacturing**

Gender employment ratio, services vs. manufacturing, selected EAP countries and comparators

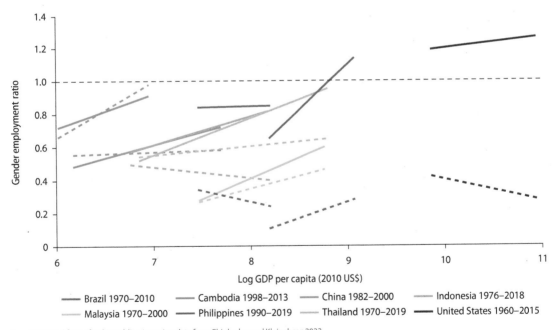

Source: Original figure for this publication using data from Chiplunkar and Kleineberg 2023.

Note: The figure plots the gender employment ratios in the services and manufacturing sectors for selected countries over time against the log of real GDP per capita in constant 2010 US$. Gender employment ratios divide the number of women working in a sector by the number of men. Solid lines are employment ratios for the services sector and dashed lines for the manufacturing sector. The covered period spans from 1960 to 2019 and varies depending on the data availability of each country. The horizontal dimension of the figure shows how fast each country's GDP per capita increased during the sample period. Data come from IPUMS (https://international.ipums.org/international/) and are complemented with labor force survey data from the World Bank's Global Labor Database. EAP = East Asia and Pacific; GDP = gross domestic product.

FIGURE O.4 **The virtuous cycle of opportunity and capacity: An organizing framework**

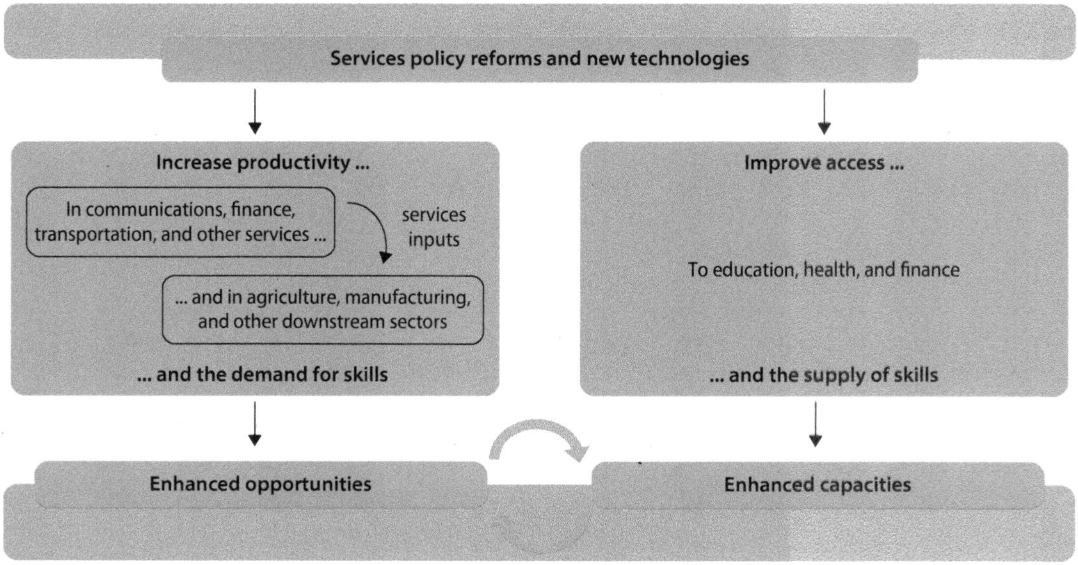

Services policy reforms and new technologies

Increase productivity ...

In communications, finance, transportation, and other services ... → services inputs

... and in agriculture, manufacturing, and other downstream sectors

... and the demand for skills

Improve access ...

To education, health, and finance

... and the supply of skills

Enhanced opportunities

Enhanced capacities

Source: Original figure for this publication.

as well as increased demand for the sophisticated skills that complement the new technologies. The same combination of services policy reform and technological diffusion is also improving education, health, and finance to equip people to take advantage of these new opportunities. However, fully unleashing the virtuous cycle between opportunity and capacity that constitutes development, as represented in the organizing framework, will require deeper reform.

The first piece of evidence for that central argument involves the growing penetration of digital technologies in a range of EAP services sectors and the link with productivity (as illustrated on the left branch of figure O.4). New firm-level evidence from the Philippines suggests that the average services firm has about a third more data and software assets per worker than its manufacturing counterpart, although the adoption of digital technologies varies across services sectors (figure O.5). Firms with access to broadband and foreign-owned firms have stronger digital technology adoption, and adoption of digital technologies is associated with higher productivity and value added within firms.

Second, evidence for EAP and other economies confirms that reducing barriers to competition in services spurs higher productivity growth in services sectors as well as in the manufacturing sectors that use those services. For example, new firm-level analysis for Viet Nam reveals that the reduction in restrictions on transportation, finance, and business sectors over the period 2008–16 was

FIGURE O.5   Services firms—especially when foreign-owned and with access to broadband—use more digital technologies, which are associated with higher levels of productivity

Services firms' data and software assets, use of data technology, and productivity and value added, the Philippines, 2017

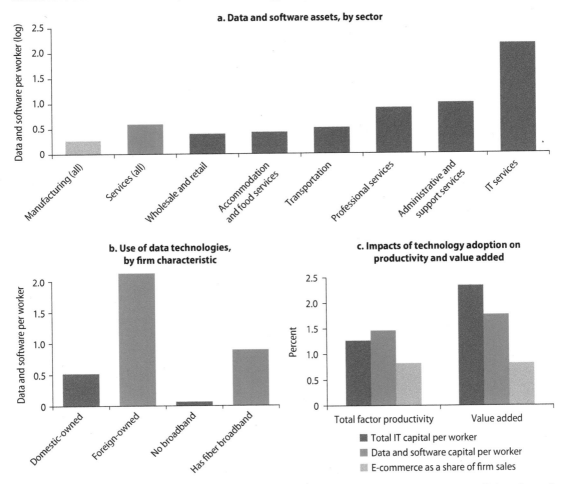

Sources: Original figure for this publication using data from the Philippines Statistical Authority: Annual Survey of Philippine Business and Industry, Census of Philippine Business and Industry, and Survey on Information and Communication Technology.
Note: IT = information technology.

associated with a 2.9 percent annualized increase in value added per worker in these sectors (figure O.6). Furthermore, the liberalization in services was associated with a 3.1 percent increase in labor productivity of the manufacturing enterprises that use services inputs, benefiting small and medium private enterprises most significantly.

FIGURE O.6   **Services liberalization increased firms' labor productivity in both services and downstream manufacturing firms**

Direct and indirect effects of services trade liberalization, by type of services firm, Viet Nam, 2008–16

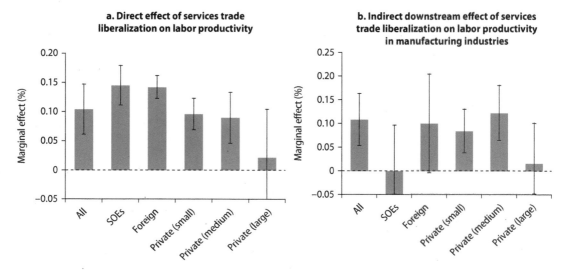

Source: Original figure for this publication using data from Viet Nam Enterprise Surveys, 2008 and 2016.
Note: Figure shows ordinary least squares regression results. The dependent variable is the change in log value added per worker between 2008 and 2016. The main explanatory variable is the change in STRI values in trade, transportation, finance, professionals, and telecommunication sectors between 2008 and 2016 in panel a, and the change in the "downstream" STRI for manufacturing sectors in panel b. The downstream STRI is a sector-specific measure for each 2-digit manufacturing sector, calculated by the average STRI of the specified five services sectors weighted by the corresponding purchasing value from each manufacturing sector. The regression sample in panel a consists of all enterprises operating in trade, transportation, finance, professionals, and telecommunication sectors, and all manufacturing enterprises in panel b, in 2008 and 2016. All regressions control for firms' baseline revenue and employment. Standard errors clustered at the industry level. SOEs = state-owned enterprises; STRI = Services Trade Restrictiveness Index.

New jobs created in digital services may, however, require greater skills than those in traditional services, as indicated by new microlevel evidence from Indonesia. Digital employment represents a larger share in services than in agriculture and manufacturing, and an even higher share in more technical services, such as information and communication technology and financial services, than in less technical sectors like distribution and transportation services (figure O.7). Furthermore, close to 40 percent of formal digital workers have a university degree or above, whereas less than 20 percent of nondigital workers do so. These findings imply a likely increase in the demand for skilled workers in the coming years. Other evidence suggests an important distinction, however: web-based digital platform jobs—delivering services like customized software and programming over the internet—tend to require higher levels of education than do location-based digital jobs—for example, ride sharing and food delivery—which are dominated by less educated workers.

Because the promise of higher productivity levels and growth is likely to be more strongly associated with more highly skilled jobs, equipping workers with the

FIGURE O.7 **In Indonesia, digital jobs often require a higher level of education and dominate in more technical services sectors**

Digital workers, by education level and sector, 2022

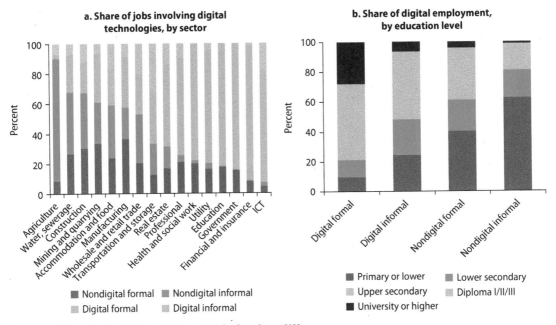

a. Share of jobs involving digital technologies, by sector

b. Share of digital employment, by education level

Legend:
- ■ Nondigital formal
- ■ Nondigital informal
- ▦ Digital formal
- ▨ Digital informal

- ■ Primary or lower
- ■ Lower secondary
- ▦ Upper secondary
- ▨ Diploma I/II/III
- ■ University or higher

Source: Original figure for this publication using Indonesia's Labor Force Survey, 2022.
Note: Digital workers are defined as workers who use digital technologies and the internet for work in their primary job. ICT = information and communication technology.

relevant skills must be a priority. The use of new digital technologies and reforms in education and health services could help address the skills deficit as well as the inequality of access and quality across the region (figure O.8), and it could equip more EAP citizens to engage productively in the new digital economy (as illustrated on the right branch of figure O.4).

The role of policy

To unleash the virtuous cycle between opportunity and capacity, and to ensure inclusive and sustainable services development, EAP countries must take three pairs of policy actions.

First, countries must pursue both liberalization and regulation. New data reveal that services trade liberalization is still unfinished business. Figure O.9 reports values for the World Bank–World Trade Organization Services Trade Restrictiveness Index by the economies' level of development. Despite past reforms, EAP countries still

FIGURE O.8 **Existing studies show that digital technologies improve education outcomes when complemented by education reform**

Size of effect of education technology programs on student learning, by type of treatment

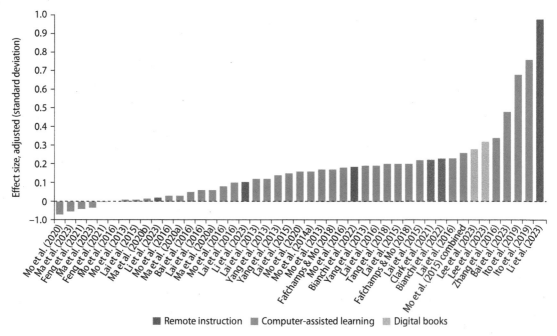

Source: Yarrow et al., annex 1, forthcoming. See annex 1 for a complete list of the studies and effects sizes (adjusted).

FIGURE O.9 **Most EAP countries restrict services trade more than other countries at comparable levels of development**

STRI values in selected EAP countries, by GDP per capita, 2022

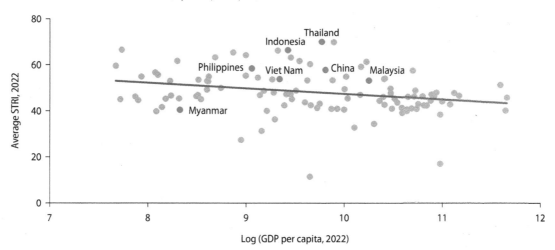

Source: World Bank, World Development Indicators, 2022; World Bank–World Trade Organization Services Trade Restrictiveness Index, 2022.
Note: The average STRI is computed as a simple average of the indicators for the financial, communication, and transportation sectors. EAP = East Asia and Pacific; GDP = gross domestic product; STRI = Services Trade Restrictiveness Index.

have relatively restrictive regimes for services. Advancing the liberalization agenda requires addressing policy restrictions on entry and competition in EAP services markets, ranging from discretionary and opaque licensing to limits on foreign ownership. In parallel, countries need to institute a regulatory framework that addresses old and new market distortions, including the concentration and data misuse that can arise in markets dominated by digital platforms.

Second, governments need to work with the private sector on building the infrastructure and skills needed to take advantage of emerging opportunities. Over the past decades, the democratization of access to mobile telephony provided by competing private firms seemed to have obviated the need for the fixed-line networks created by plodding public sector monopolies (table O.1). However, the digital benefits of access to high-speed broadband have revived the question of how the state can ensure adequate access for the poor and remote populations. Competitively allocated subsidies to private providers may help bridge the gaps, as the experience with universal access funds for basic telecommunications shows. Countries must also wrestle with the question of how far the market and private institutions can be relied on to deliver the skills needed by the digital services economy (figure O.10). At the very least, governments can remedy inadequacy and inequality in access to finance for education, ensure the quality of educational institutions without impeding competition, and address coordination failures by ensuring coherence in policies and providing information to individuals and firms on the evolution of economic opportunities and human capacities.

FIGURE O.10   Tertiary enrollment rates have increased in the EAP region but remain low in most countries, with the private sector playing a limited role

Tertiary education enrollment, total and private, selected EAP countries

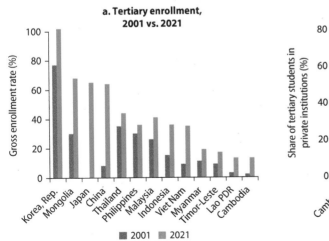
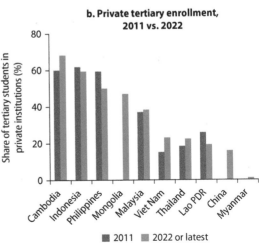

Source: International Labour Organization Department of Statistics, ILOSTAT, 2022.
Note: EAP = East Asia and Pacific.

TABLE O.1 Mobile broadband access has improved in most countries, but mobile does not offer the potential speed of fixed broadband, access to which remains limited and unequal

Access, quality, and cost of fixed and mobile broadband, EAP countries

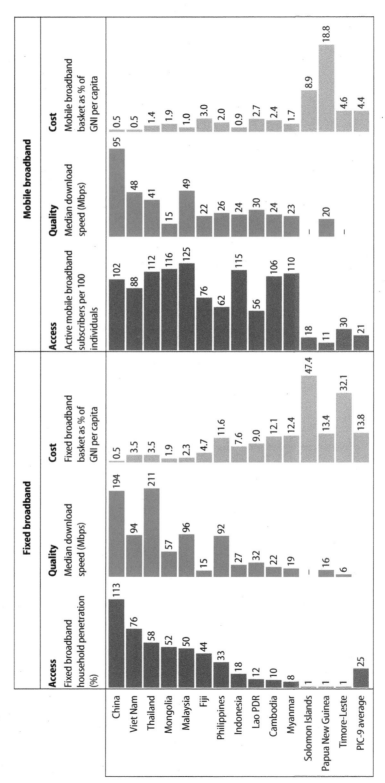

	Fixed broadband			Mobile broadband		
	Access Fixed broadband household penetration (%)	**Quality** Median download speed (Mbps)	**Cost** Fixed broadband basket as % of GNI per capita	**Access** Active mobile broadband subscribers per 100 individuals	**Quality** Median download speed (Mbps)	**Cost** Mobile broadband basket as % of GNI per capita
China	113	194	0.5	102	95	0.5
Viet Nam	76	94	3.5	88	48	0.5
Thailand	58	211	3.5	112	41	1.4
Mongolia	52	57	1.9	116	15	1.9
Malaysia	50	96	2.3	125	49	1.0
Fiji	44	15	4.7	76	22	3.0
Philippines	33	92	11.6	62	26	2.0
Indonesia	18	27	7.6	115	24	0.9
Lao PDR	12	32	9.0	56	30	2.7
Cambodia	10	22	12.1	106	24	2.4
Myanmar	8	19	12.4	110	23	1.7
Solomon Islands	1	–	47.4	18	–	8.9
Papua New Guinea	1	16	13.4	11	20	18.8
Timore-Leste	1	6	32.1	30	–	4.6
PIC-9 average	25		13.8	21		4.4

Sources: International Telecommunication Union; Ookla; TeleGeography.

Note: The table shows numerical average of the available data points for PIC-9. EAP = East Asia and Pacific; GNI = gross national income; Mbps = megabits per second; PIC-9 = Kiribati, Marshall Islands, Micronesia, Nauru, Palau, Samoa, Tonga, Tuvalu, and Vanuatu; – = not available.

Third, EAP and other countries must complement unilateral domestic reform with cooperative international action to address services market failures that have a transborder dimension. One example is the need to ensure that heterogeneity in national regulatory approaches to privacy and cybersecurity (as shown in map O.1) do not impede the data flows central to the global services economy. Another example is the need to ensure that international transportation, central to global trade and tourism, does not continue to add carbon dioxide emissions to the atmosphere. In both these cases, countries are beginning to cooperate meaningfully in regional and multilateral forums.

Conclusions

Looking ahead, it will be important for countries in the EAP region to pursue development strategies that are attuned to the interplay between opportunity and capacity. As illustrated in figure O.11, favoring the development of opportunities without the accompanying increased capacities could lead to potential shortages. Developing capacities in the absence of opportunities could generate potential underutilizations. A virtuous cycle that powers development can only be achieved by pursuing balanced policies, in which the enhancement of endowments shapes comparative advantage, and the evolution of comparative advantage incentivizes the enhancement of endowments.

MAP O.1   EAP countries take a heterogeneous regulatory approach to international data transfers

Cooperation in cross-border data flows, by type of approach

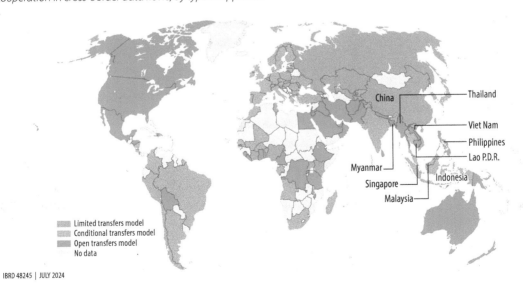

IBRD 48245 | JULY 2024

Source: World Bank 2021, updated for EAP-6 countries in June 2023.
Note: EAP = East Asia and Pacific.

FIGURE O.11   **The need for a balanced development of opportunities and capacities**

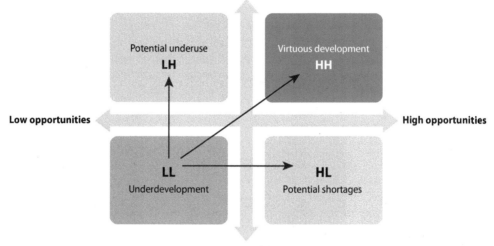

Source: Original figure for this publication using data from the World Bank.

References

Chiplunkar, G., and T. Kleineberg. 2023. "Gender Barriers, Structural Transformation and Economic Development." Working paper. https://drive.google.com/file/d/1pL9LXx6LCAR9acFcLnK_CEFJkKihRaUn/view.

Kruse, H., E. Mensah, K. Sen, and G. J. de Vries. 2023. "A Manufacturing (Re)naissance? Industrialization Trends in the Developing World." *IMF Economic Review* 71: 439–73.

World Bank. 2021. *World Development Report 2021: Data for Better Lives.* Washington, DC: World Bank.

Yarrow, N., C. Abbey, S. Shen, and K. Alyono. Forthcoming. *Using Education Technology to Improve K–12 Student Learning in East Asia and Pacific: Promises and Limitations.* Washington, DC: World Bank.

Introduction

1

Abstract

The East Asia and Pacific (EAP) region's rapid economic growth in recent decades is seen as manufacturing-led, but services are important yet underappreciated drivers of growth and job creation. Although services are often seen as slow to innovate, tough to trade, and hard to liberate from regulatory restrictions, a combination of technological change and policy reform is transforming services into the most dynamic part of many economies. In fact, the development of services will be central to the EAP region's overall development.

Seven facts about services in East Asia and Pacific

Let's start considering seven facts about services in the EAP region (refer to box 1.1 for a discussion of how to define services). First, from 2010 to 2021, both China and the rest of the EAP region experienced a significant increase in the share of services and a decrease in the share of manufacturing in value added (figure 1.1). The share of services in employment across the region also increased. Although the share of manufacturing in employment decreased in China, it increased in other EAP countries, most notably Cambodia and Viet Nam (refer to appendix A, which provides a breakdown for other EAP economies).

Second, in the larger EAP countries, the increase in employment in services has been mostly directed to traditional, lower-productivity sectors, such as construction and trade, rather than higher-productivity sectors, such as business services, finance, and transportation (figure 1.2, panel a). Nevertheless, on aggregate, the services sector has been a consistent and resilient contributor to aggregate labor productivity growth. Moreover, in the most recent years, the contribution of the services sector to aggregate labor productivity growth has been higher than the contribution of manufacturing in all economies (figure 1.2, panel b).

Third, services currently represent the most dynamic elements of international trade and foreign direct investment (FDI) in East Asia (figure 1.3, panels a and b).

Box 1.1.   What are services?

Somewhat surprisingly, there is no single definition of "services." The alternative definitions are based on different attributes of services: intangibility, nonstorability, simultaneity of production and consumption, and so on. These definitions are not watertight: customized software on a disk is storable, produced before it is consumed, and tangible (at least the disk, if not the service itself). Instead of trying to converge on a precise definition, it is convenient to identify services by relying on an open-ended classification of services, such as the following one used by the World Trade Organization:

1. Business services
2. Communication services
3. Construction services
4. Distribution services
5. Educational services
6. Environmental services
7. Financial services
8. Health-related and social services
9. Tourism and travel-related services
10. Recreational, cultural, and sporting services
11. Transportation services
12. Other services not elsewhere included.

Other classifications are naturally possible. For instance, refer to Nayyar, Hallward-Driemeier, and Davies (2021) for a classification based on tradability and skill intensities.

FIGURE 1.1   Share of services in value added and employment increased significantly in recent years

Value added and employment of manufacturing and services, China and regional comparators, 2010 and 2021

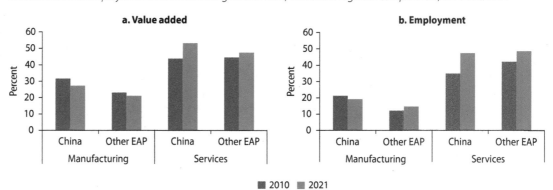

Sources: International Labour Organization Department of Statistics, ILOSTAT; World Bank, World Development Indicators.
Note: EAP = East Asia and Pacific.

FIGURE 1.2   On aggregate, services are important contributors to productivity growth

Contribution of services to labor productivity and employment, selected countries

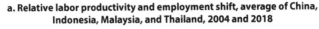

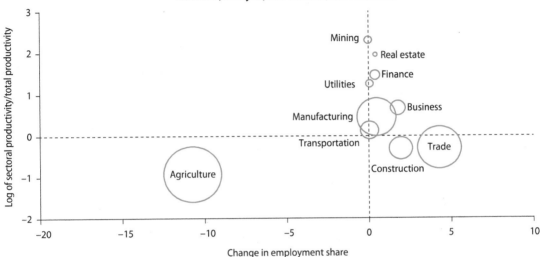

a. Relative labor productivity and employment shift, average of China, Indonesia, Malaysia, and Thailand, 2004 and 2018

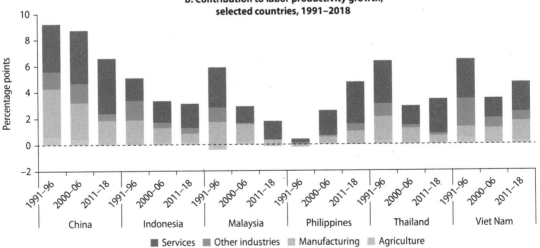

b. Contribution to labor productivity growth, selected countries, 1991–2018

Source: Groningen Growth and Development Centre and United Nations, GGDC/UNU-WIDER Economic Transformation Database (Kruse et. al. 2023).
Note: Panel a is based on productivity and employment averages for China, Indonesia, Malaysia, and Thailand. The size of the circle representing each sector reflects the employment share of that sector in 2004. The horizontal axis measures the change in employment share from 2004 to 2018, showing a shift from agriculture to services. The vertical axis represents the relative productivity of the different sectors (relative to overall average productivity), from which it is evident that most of the increase in employment has been absorbed by relatively low-productivity sectors. Panel b shows sectoral contribution to labor productivity growth, annual average during three different periods. Other industries include mining, utilities, and construction.

FIGURE 1.3   **Services outpaced manufacturing in terms of growth of trade and FDI in recent years**

Exports and FDI growth, selected EAP countries

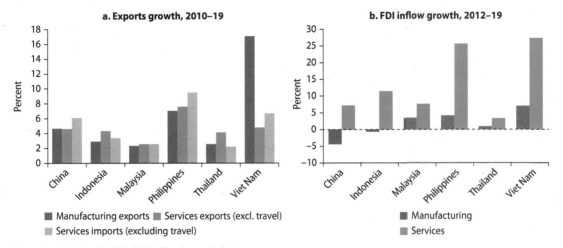

Sources: Haver Analytics; World Bank, World Development Indicators.
Note: Panel a reports the annual growth in the export and import of services for several EAP countries. For this analysis, the services sector excludes travel. The growth rate of manufacturing exports is included for reference. Panel b shows average annual growth of three-year moving-average FDI values, during 2012–19, except for Viet Nam (2014–19) and Indonesia and the Philippines (2013–19). EAP = East Asia and Pacific; FDI = foreign direct investment.

In all the countries presented in figure 1.3, panel a, except for Viet Nam, the rate of growth of services exports is higher than for manufacturing. Strikingly, in all countries considered in figure 1.3, panel b, the growth rate of FDI inflows in services vastly outpaced the growth of FDI in manufacturing.[1] Moreover, the increase in the trade of digitally deliverable services (for instance, financial services, telecommunications, and computer and information services) over the 2005–21 period is almost double the increase in the trade of goods (figure 1.4).[2]

Fourth, demand for services will likely expand at a faster pace with the rapidly growing incomes and emerging middle class across the region. Over the past decade, per capita incomes for lower- and middle-income countries in the EAP region have grown by over 10 percent a year on average. As shown in figure 1.5, evidence from Indonesia and the Philippines confirms that more affluent households tend to spend a higher share on the consumption of services, especially in sectors like health, education, and communications. Based on this evidence, one can expect an increase in the demand for services as the levels of income rise across the region.

Fifth, on the labor market side, services tend to employ more highly skilled workers than manufacturing or agriculture. This statement is supported by labor market evidence from Viet Nam, where 40 percent of the people working in the services sectors in 2019 had a tertiary education degree. The corresponding numbers were 10 percent for manufacturing and 1 percent for agriculture (figure 1.6).

FIGURE 1.4   Trade in digitally delivered services increased more than in other sectors in recent years

Trade in services, selected countries, 2005–21

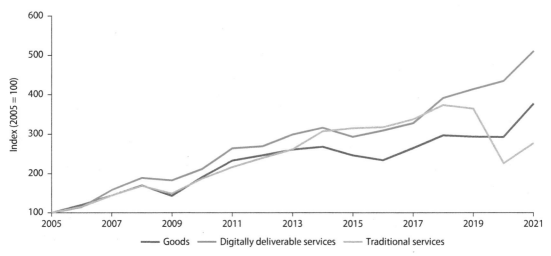

Source: United Nations Conference on Trade and Development.
Note: Digitally deliverable services are an aggregation of insurance and pension services, financial services, charges for the use of intellectual property, telecommunications, computer and information services, other business services, and audiovisual and related services. The figure includes Cambodia, China, Fiji, Indonesia, Kiribati, Malaysia, Mongolia, Papua New Guinea, the Philippines, the Solomon Islands, Thailand, Tonga, Vanuatu, and Viet Nam. Trade is defined as the sum of imports and exports.

FIGURE 1.5   Demand for services rises with income level

Share of total expenditure, by sector and income percentile, Indonesia and the Philippines, 2021

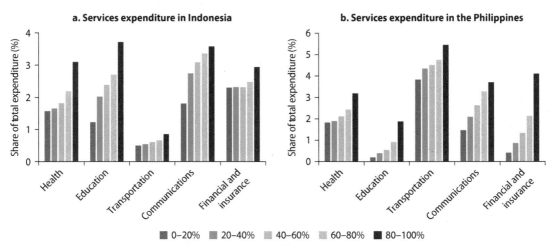

Sources: Indonesian National Socioeconomic Survey and Family Income and Expenditure Survey in the Philippines, 2021.

FIGURE 1.6   Structural transformations toward services shift demand to more highly educated workers

Share of employment, by sector and education level, Viet Nam, 2019

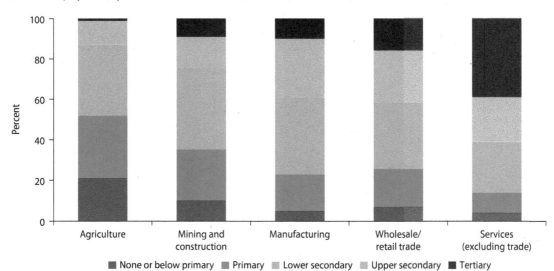

Source: Viet Nam Labor Force Survey, 2021.

Sixth, the services sector has higher female employment shares than manufacturing. Figure 1.7 shows the ratio of female to male workers in the services sector (solid line) and the manufacturing sector (dashed line) in various EAP countries and comparators. Complete gender equality would imply a ratio of 1 in all sectors. Over time, gender employment ratios in EAP countries increased rapidly in the services sector but stagnated or even decreased in the manufacturing sector. Overall, the expected future path of structural transformation and the rise of services are likely to improve gender equality and gender representation in the labor market, thus stimulating gender-inclusive growth.

Last, as shown in figure 1.8, services sectors, except transportation, are characterized by a lower emission intensity than agriculture and manufacturing in both China and Indonesia. Therefore, structural transformation toward services would support the region's shift to lower carbon growth. This result applies to both direct and indirect emissions.

An organizing framework and the rest of this report

Figure 1.9 presents the organizing framework of this report. At the top appear two major drivers that have shaped and will influence the evolution of services and their future contribution to development: technological change and services policy reforms. Despite the uneven diffusion of digital technologies and the limited reform of policies restricting entry and competition in services sectors, those sectors are leading to structural change across and within sectors. The result, through increased

FIGURE 1.7 **The services sector has higher female employment shares than manufacturing**

Gender employment ratios in manufacturing and services, selected countries

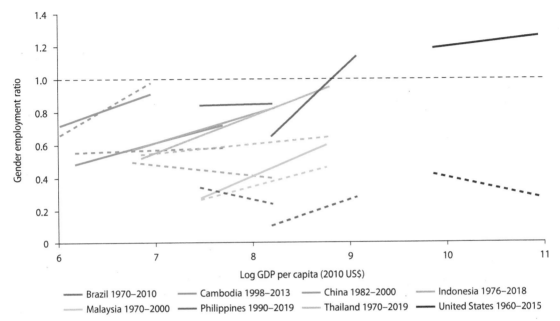

Source: Original figure for this publication using data from Chiplunkar and Kleineberg 2023.
Note: The figure plots gender employment ratios in the services and manufacturing sectors for selected countries over time against the log of real GDP per capita in constant 2010 US dollars. Gender employment ratios divide the number of women working in a sector by the number of men. Solid lines are employment ratios for the services sector and dashed lines for the manufacturing sector. The covered period spans from 1960 to 2019 and varies depending on the data availability of each country. The horizontal dimension of the figure shows how fast each country's GDP per capita increased during the sample period. Data come from IPUMS International (https://international.ipums.org/international/), complemented with labor force survey data from the World Bank's Global Labor Database. GDP = gross domestic product.

scale, tradability, and innovation, is higher productivity growth in services sectors—as well as in the manufacturing sectors that use these services—and increased demand for the sophisticated skills that complement the new technologies. As represented in the left branch of the figure 1.9, these developments generate enhanced economic opportunities. The same combination of services policy reforms and technological diffusion can also improve access to education, health, and finance to equip people to take advantage of these new opportunities by increasing the supply of skills (right branch), thus creating enhanced capacities. The last element (at the bottom) shows the possibility of a virtuous cycle between opportunities and capacities, which require the right set of policies.

The rest of this report is structured as follows. Chapter 2 ("What Is Happening?") explores the top panel of the organizing framework, namely the diffusion of new technologies (with a particular emphasis on digitalization) and services policy reforms. Chapter 3 ("Why Do Digitalization and Service Reforms Matter?") deals with the middle part of the framework. It first explores the left branch, namely the impact of the key drivers on productivity and the demand for skilled jobs, then the

FIGURE 1.8 **Services are generally less carbon intensive than agriculture and manufacturing (except in transportation)**

Greenhouse gas emission intensity, by sector, China and Indonesia

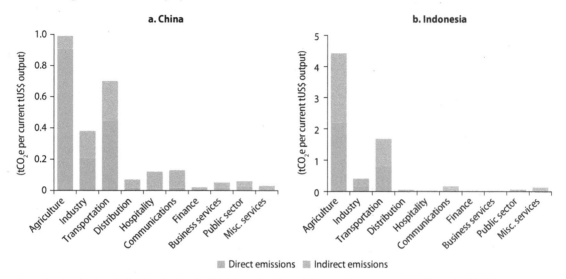

Sources: Based on data from the Emissions Database for Global Atmospheric Research (European Commission and JRC/PBL 2016) and Global Resource Input-Output Assessment (Lenzen et al. 2017) databases. Emissions include carbon dioxide (CO_2), methane (CH_4), and nitrous oxide (N_2O). tCO_2e = tons of carbon dioxide equivalent.

FIGURE 1.9 **The virtuous cycle of opportunity and capacity: An organizing framework**

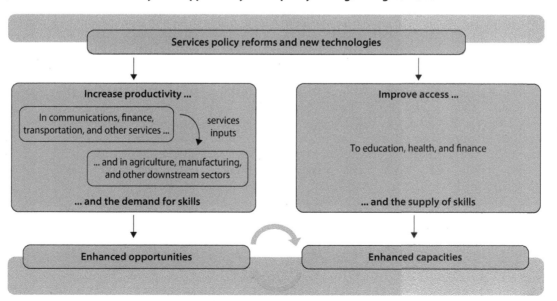

Source: Original figure for this publication.

right branch, investigating the impact of the key drivers on access to education, health, and finance. Chapter 4 ("What Needs to Be Done?") focuses on the policies needed to unleash a virtuous cycle of opportunities and capacities, represented at the bottom of the organizing framework. The final chapter concludes, and appendixes provide supporting information.

Notes

1. Appendix B reports the sectoral disaggregation of FDI inflows in services for some EAP countries.
2. Whereas figure 1.4 reports an aggregate dynamic for the entire EAP region, appendix C reports results for individual countries. Refer also to ADB (2022) for an in-depth study of digital trade in the EAP region.

References

ADB (Asian Development Bank). 2022. *Unlocking the Potential of Digital Services Trade in Asia and the Pacific*. Manila: ADB.

Chiplunkar, G., and T. Kleineberg. 2023. "Gender Barriers, Structural Transformation and Economic Development." Unpublished manuscript. https://drive.google.com /file/d/1pL9LXx6LCAR9acFcLnK_CEFJkKihRaUn/view.

European Commission and JRC (Joint Research Centre)/PBL (Netherlands Environmental Assessment Agency). 2016. "Emission Database for Global Atmospheric Research (EDGAR)," release version 4.3.1. https://edgar.jrc.ec.europa.eu/dataset_tox4.

Kruse, H., E. Mensah, K. Sen, and G. J. de Vries. 2023. "A Manufacturing (Re)naissance? Industrialization Trends in the Developing World." *IMF Economic Review* 71: 439–73.

Lenzen, M., A. Geschke, M. D. Abd Rahman, Y. Xiao, J. Fry, R. Reyes, E. Dietzenbacher, S. Inomata, K. Kanemoto, B. Los, D. Moran, H. Schulte in den Bäumen, A. Tukker, T. Walmsley, T. Wiedmann, R. Wood, and N. Yamano. 2017. "The Global MRIO Lab—Charting the World Economy." *Economic Systems Research* 29 (2): 158–86.

Nayyar, G., M. Hallward-Driemeier, and E. Davies. 2021. *At Your Service? The Promise of Services-Led Development*. Washington, DC: World Bank.

What Is Happening?   2

Abstract

This chapter introduces two important factors affecting services sectors in the East Asia and Pacific (EAP) region: digitalization and policy reforms. Developments in each area have reshaped the nature of services with consequences for productivity, jobs, and access.

Significant but uneven digitalization

In recent years, countries in EAP have digitalized rapidly, primarily on the back of evolving mobile and internet technologies. The share of the region's population using the internet jumped from 10 percent in the early 2000s to over 70 percent in 2021 (ITU 2022). Beyond the household level, digitalization has also reshaped how firms do business across services sectors as well as in agriculture and industry. The pace of digital technology adoption has been rapid and exceptionally transformative in services delivery, allowing firms to provide services faster, cheaply, and across multiple markets. According to the global *Digital Progress and Trends Report 2023*, EAP has been ahead of other developing regions in business digitalization, with the share of firms investing in digital solutions quadrupling from 13 percent to 54 percent between 2020 and 2022 (World Bank 2024). In other regions, less than 30 percent of firms had made similar investments by the end of 2022. The rapid growth of digital platforms is worth noting, particularly in smaller EAP economies. The platform economy in the EAP region has grown by an estimated 5–7 percent of gross domestic product (GDP) in under 10 years, with the fastest growth occurring in Indonesia, Thailand, and Viet Nam. However, empirical analysis of firms in the Philippines (detailed in the following subsection) shows that, despite the productivity-enhancing benefits of digitalization, adoption has been uneven both across and within sectors. Furthermore, funding for digital services firms in the region is skewed—with China, Japan, and the Republic of Korea dominating in the number of digital businesses and amount of venture capital funding.

Digitalized businesses

This subsection looks at the digitalization of businesses using three different lenses: (1) technology adoption, (2) the flow of venture capital investments, and (3) the diffusion of digital platforms.

Firms in the EAP region have increasingly adopted digital technologies, especially in the services sector. In the case of the Philippines, matching detailed and representative technology data for both manufacturing and services firms with firm-level productivity surveys provides empirical evidence of this phenomenon. The data show that the digital intensity of Philippine services firms is higher than for manufacturing firms, as measured by their investments in information technology capital per worker (table 2.1). Moreover, the services sector had greater growth of digital technology adoption between 2012 and 2017, particularly for technologies related to data, such as cloud computing, data analytics, and investments in databases and software.

Despite overall increased adoption of technologies, the extent of digitalization within firms varies enormously across services subsectors (figure 2.1). On average, Philippine services firms have about a third more data and software assets per worker than their manufacturing counterparts, but this average masks substantial sectoral differences. Technology services firms have, on average, seven times more data and software assets per worker than manufacturers do.

Furthermore, use of data is uneven across services firms in the Philippines. Foreign firms are five times more data-intensive than domestically owned firms (figure 2.2). Firms with fiber are more than twice as data-intensive as those without broadband, but foreign firms with fiber are more than 15 times more data-intensive than domestic firms without it.

TABLE 2.1 Technology adoption in the Philippines is higher in services than in manufacturing
Technology adoption by firms in the Philippines, 2012 and 2017

	Manufacturing		Services	
	2012	2017	2012	2017
Total IT capital per worker (log)	2.81	2.77	3.59	4
Databases and software capital per worker (log)	0.17	0.27	0.36	0.62
Cloud computing (% of firms)	—	18	—	38
Devote IT resources to data or analytics (% of firms)	5	30	8	42
E-commerce sales (% of firms)	1	25	0	17
Share of sales via e-commerce (% of firm sales)	0.30	4.90	0.10	1.00

Sources: Philippines Statistical Authority: Annual Survey of Philippine Business and Industry, Census of Philippine Business and Industry, and Survey on Information and Communication Technology.
Note: Table shows representative statistics for business sector services firms (ISIC rev 4 divisions 45–82) and manufacturing firms (ISIC rev 4 divisions 10–33) with 20 or more employees. To avoid dropping zero values, the analysis added one Philippine peso per worker to capital before taking logs. ISIC = International Standard Industrial Classification of All Economic Activities; IT = information technology; — = not available.

FIGURE 2.1   Use of data technologies varies across services sectors in the Philippines

Services firms' data and software assets, by sector, 2017

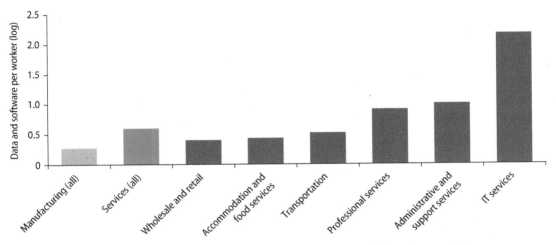

Sources: Philippines Statistical Authority: Annual Survey of Philippine Business and Industry, and Census of Philippine Business and Industry.
Note: Figure shows representative statistics for business sector services firms (ISIC rev 4 divisions 45-82) with 20 or more employees in 2017 and reflects log data and software capital per worker (PPP, 2005 international dollars). To avoid dropping zero values, the analysis added one Philippine peso per worker to data and software capital before taking logs. ISIC = International Standard Industrial Classification of All Economic Activities; IT = information technology; PPP = purchasing power parity.

FIGURE 2.2   FDI plays a crucial role in the diffusion of data technologies in Philippine services firms, especially when complemented with high-speed broadband

Services firms' use of data technologies, by firm characteristics, the Philippines, 2017

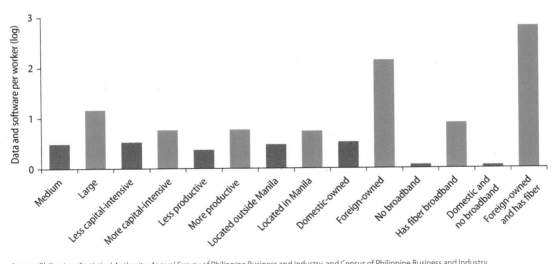

Sources: Philippines Statistical Authority: Annual Survey of Philippine Business and Industry, and Census of Philippine Business and Industry.
Note: Large refers to firms with 100 or more employees, and medium refers to firms with between 20 and 29. More capital-intensive (more productive) reflects firms with capital per worker (total factor productivity) above the median within their two-digit industry, with those below the median defined as less-capital intensive (less productive). Foreign-owned reflects at least 50 percent foreign ownership, with the remainder being domestically owned. No broadband reflects either narrowband internet access or no internet access. Figure shows representative statistics for business sector services firms (ISIC rev 4 divisions 45-82) with 20 or more employees in 2017 and reflects log data and software capital per worker (PPP, 2005 international dollars). To avoid dropping zero values, the analysis added one Philippine peso per worker to data and software capital before taking logs. FDI = foreign direct investment; ISIC = International Standard Industrial Classification of All Economic Activities; PPP = purchasing power parity.

FIGURE 2.3 **China dominates the region in number of digital businesses and amount of venture capital funding**

Digital businesses and venture capital funding, EAP region, 2022

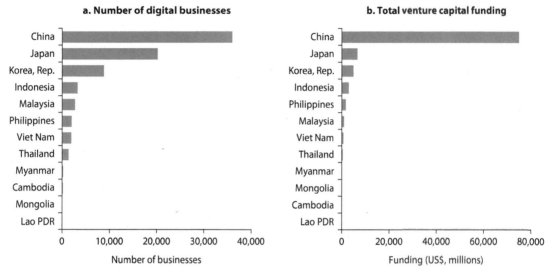

a. Number of digital businesses

b. Total venture capital funding

Source: FCI Digital Business Database.
Note: EAP = East Asia and Pacific.

Looking beyond technology adoption within firms and focusing on technology providers shows that, as expected, China, EAP's largest economy, dominates the landscape of digital businesses and venture capital funding in the region (figure 2.3). This analysis draws upon a newly assembled firm-level database: the FCI Digital Business Database (described in Zhu et al. 2022). The database covers about a million digital businesses (founded between 1970 and 2022) in 200 countries and funding flows to digital businesses as of 2022 (mostly early-stage risk financing to start-ups).

A strong positive correlation also exists between the level of development and both the number of digital businesses and the level of funding, as shown in analyses plotting those two outcomes as a function of the level of development—and controlling for the impact of country size (figure 2.4). The positions of EAP countries in terms of number of digital businesses roughly align with the levels predicted by their level of development, whereas they do slightly better than predicted in terms of funding.

The allocation of venture capital funding across different types of digital services suggests both similarities and differences in the development of the digital economy within EAP, and between EAP and other regions. For instance, financial technology and e-commerce are among the highest recipients of funding in both EAP and developed economies. By contrast, the share of funding going to health technology in developed economies is roughly double that observed in the EAP region, where mobility technology instead appears to be a more funded sector (figure 2.5; refer to appendix D for precise definitions of the sectors). Zooming in on specific countries, the analysis compared the shares of funding across different sectors in China, Indonesia, Malaysia, and Thailand (refer to appendix E). The clearest pattern

FIGURE 2.4   Most EAP countries do slightly better in terms of funding than predicted by their level of development, controlling for size

Correlation between income level and number of digital businesses and total VC funding, EAP countries, 2022

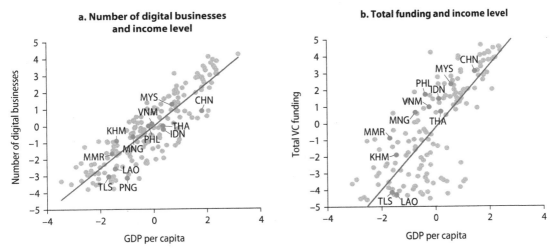

Source: FCI Digital Business Database.
Note: The scatterplots reported in figure 2.4 are avplots, representing correlations between the number of digital businesses and total funding with income levels, conditional on the level of the population. The panels represent the correlation between the residuals of two separate regressions. Panel a shows regression of both the number of digital businesses (in logs) and the income level on the population (in logs). It then plots for each country the two residuals (whose units are the number of digital businesses (in logs), as well as the best linear fit. Panel b repeats the same procedure, substituting the number of digital businesses (in logs) with the total funding (in log dollars). EAP = East Asia and Pacific; VC = venture capital.

is that funding to digital firms is much more concentrated in Indonesia, Malaysia, and Thailand than in China, whose digital economy is unsurprisingly deeper and more diverse.

Most venture capital funding in the region so far has gone to consumer services. Categorizing services into business-to-business (B2B, such as artificial intelligence or big data analytics), business-to-consumer (B2C, such as mobility or entertainment technology), or mixed (e-commerce and financial and health technology), figure 2.6 shows some similarities and an interesting asymmetry between developed and developing countries. Venture capital funding goes more to B2B sectors than to B2C sectors in developed countries, whereas the opposite pattern is observed in developing countries. Although, as already noted, financial technology and e-commerce dominate both groups of countries, this asymmetry is interesting because, arguably, B2B digital businesses are relatively more relevant for productivity gains and B2C businesses for market access (though B2B digital businesses can also foster market access).

Finally, EAP accounts for a third of the world's largest digital platform firms, second only to the United States (figure 2.7, panel a). In 2012, Asia was home to only two unicorns (privately held start-ups valued at US$1 billion or more); as of 2022, the region had 185 unicorns. The combined valuation of these firms has grown to about US$780 billion, the equivalent of 3.7 percent of the combined GDP of countries from which they originate, up from virtually zero a decade ago (figure 2.7, panel b). Some of EAP's most valuable and fastest-growing firms—China's Alibaba,

FIGURE 2.5 **E-commerce and financial and mobility technology attract high shares of venture funds in developing EAP and developed economies**

Venture capital funding, by type of technology, EAP and developed economies, 2022

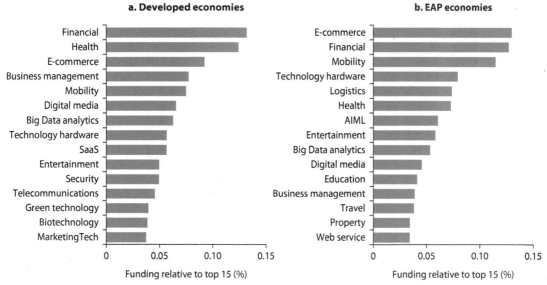

Source: FCI Digital Business Database.
Note: The bars represent the shares of each sector in the total funding going to the top-15 sectors in developed countries (panel a) and the EAP region (panel b) in 2022. Refer to appendix D for precise definitions of the sectors. AIML = artificial intelligence and machine learning; EAP = East Asia and Pacific; SaaS = Software as a Service.

FIGURE 2.6 **On average, funding goes more to B2B in developed countries and more to B2C in developing countries**

Percentage of venture capital funding, by type of business and country income level, 2022

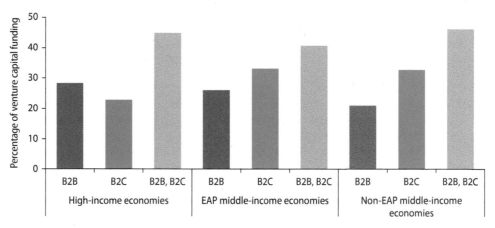

Source: FCI Digital Business Database.
Note: B2B = business-to-business; B2C = business-to-consumer; EAP = East Asia and Pacific.

FIGURE 2.7   Platform business has expanded rapidly in East Asia and Pacific
Platform business in East Asia and Pacific, compared to other regions and by industry

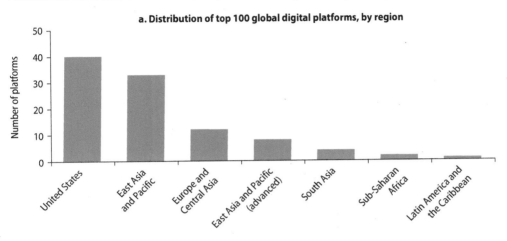

a. Distribution of top 100 global digital platforms, by region

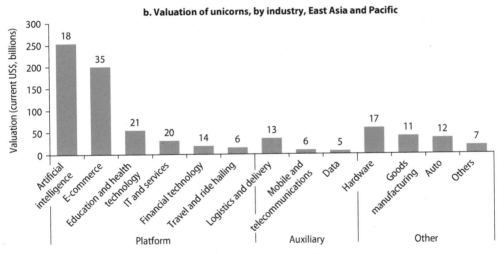

b. Valuation of unicorns, by industry, East Asia and Pacific

Sources: CB Insights (https://www.cbinsights.com/research-unicorn-companies); FactSet Fundamentals.
Note: In panel a, platforms include both public and private firms; the panel shows the regional distribution of the top 100 (globally ranked) platforms. In panel b, market capitalization is calculated using the closing prices on the last trading day of the year. The number above each bar indicates the number of unicorn firms in each industry. IT = information technology.

Meituan, and Tencent; Indonesia's GoTo; and Singapore's Grab—are major players in e-commerce, delivery, and ride hailing, with revenues that rival those of Amazon, eBay, and Uber. This dynamism extends beyond these well-known examples, however. Of the 20 largest publicly listed private sector firms in EAP, 8 are technology-based services companies with a combined market capitalization (a reflection of expected future revenue streams from these businesses) of US$1.05 trillion (5.8 percent of China's GDP in 2023).[1] Based on market capitalization, the EAP region had 4 of the world's 10 most valuable technology-based services firms in 2022, whereas it had no companies in the global top 10 in 2012.

The size of the platform economy in the EAP region, although still modest, has increased rapidly for some EAP countries. The evolution of its gross merchandise value offers a way to visualize the rapid growth over time of the platform economy. Figure 2.8 shows the estimate provided by Google, Temasek, and Bain & Company (2022) in their annual report on the digitalization of Southeast Asia (defined in that report as Indonesia, Malaysia, the Philippines, Thailand, Singapore, and Viet Nam). The figure shows rapid growth in estimated gross merchandise value between 2015 and 2022 for all the EAP countries considered. At current market rates, the size of the digital economy according to this measure is between 5 percent and 7 percent of GDP.

Digitalized households

Beyond businesses, the digital connectivity of people, especially through mobile technology, has expanded dramatically since 2012. Today, the EAP region is home to 1.5 billion active internet users, accounting for 70 percent of the total population. Growth in mobile broadband subscriptions has far outstripped fixed broadband growth since the early 2010s, with subscriptions increasing from 437 million in 2013 to 2 billion in 2021. Gaps in access remain, however, both within and across countries, as shown in figure 2.9, which reports the share of households with internet access in seven EAP countries. The figure shows large differences between countries, with all countries exhibiting three sources of digital divides. Unsurprisingly, the share of households with access to the internet is higher in urban areas, among the richest households, and among the most educated ones.[2]

FIGURE 2.8   **The size of the platform economy has increased rapidly to 5–7 percent of GDP**

Gross merchandise value of platform economy, by industry, selected EAP countries, 2015, 2019, and 2022

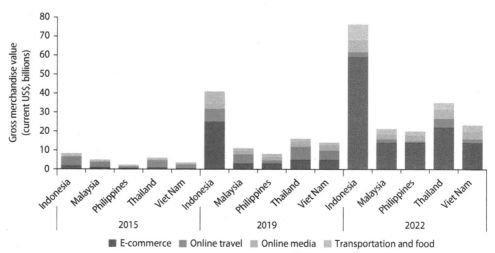

Source: Google, Temasek, and Bain & Company 2022.
Note: EAP = East Asia and Pacific.

Limited network speed presents a significant constraint in the EAP region, where download speed averages only 40 percent of that in high-income economies. Despite extensive mobile cellular coverage, the rate of high-speed mobile coverage (specifically fourth generation), averages only about 65 percent across countries. Notably, early adopters like Malaysia and Thailand reached fourth-generation penetration levels comparable to levels in advanced economies, whereas others like the Pacific Islands lag behind. Additionally, the low penetration of fixed internet technologies contributes to limited high-speed internet access. Disparities in internet coverage, quality, and cost persist across countries, with examples like the Philippines experiencing speeds more than 55 percent slower than global averages in 2020.

FIGURE 2.9 **EAP countries have unequal internet access, with strong urban-rural, income, and education divides**

Internet access, by country, place of residence, income quintile, and education level, selected EAP countries

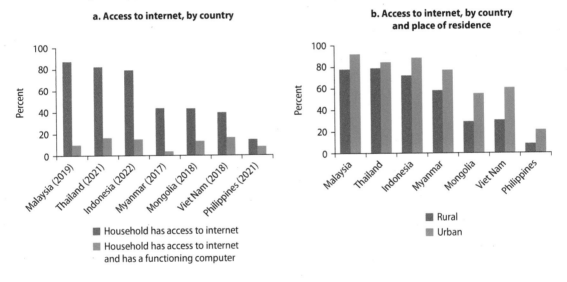

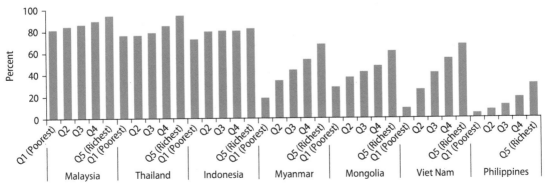

(Continued)

FIGURE 2.9   EAP countries have unequal internet access, with strong urban-rural, income, and education divides *(Continued)*

d. Access to internet, by country and education level

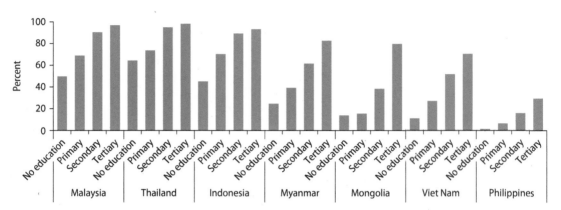

Sources: World Bank East Asia and Pacific Team for Statistical Development using harmonized survey data from Indonesia National Socio-Economic Survey, 2022; Malaysia Household Income, Expenditure and Basic Amenities Survey, 2019; Mongolia Household Socio-Economic Survey, 2018; Myanmar Living Conditions Survey, 2017; Philippines Family Income and Expenditure Survey, 2021; Thailand Household Socio-Economic Survey, 2021, and Viet Nam Household Living Standards Survey, 2020.
Note: Country statistics computed using two different internet access proxy indicators depending on survey availability: (1) "The household has an internet subscription" (Malaysia, Myanmar, Philippines, and Viet Nam), more likely capturing access to fixed internet broadband only; and (2) "The household has internet access, whether inside or outside the house" (Indonesia, Mongolia, and Thailand), more likely capturing access to both fixed and mobile internet broadband.

The evidence presented in this section leaves little doubt that digitalization has been accelerating across the region. Questions remain, however, about the depth of the impact of these developments and the breadth of the potential consequences for productivity, jobs, and development.

Significant but uneven services reforms

In addition to digitalization, structural reforms constitute a second driver affecting the services sector in the EAP region. Beginning in the early 1980s until the early 2000s, major economies in the region implemented significant structural reforms, but the pace of reforms has slowed (figure 2.10). The earlier reforms were broad-based and covered multiple sectors such as domestic finance, communications, and electricity product markets. These reforms allowed greater private participation, including through privatization of some public monopolies and elimination of restrictions on entry and operations by new domestic and foreign firms. The available evidence suggests that, in most countries, the pace of reforms slowed after the early 2000s. China and Viet Nam continued to implement reforms until later, driven in part by their accession commitments to the World Trade Organization, but even the pace of their reforms slowed starting in the mid-2010s. Recent reforms in Indonesia and the Philippines are discussed later.

To an extent, the slowing momentum of reform is a consequence of significant prior liberalization that leaves limited room for further reforms, especially the politically

FIGURE 2.10   Most EAP countries implemented significant reforms until the early 2000s but have introduced fewer since then

Reforms in selected EAP countries, by index, 1970s to 2020

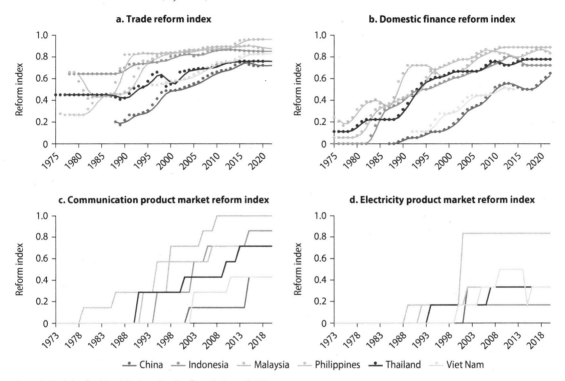

Source: Original data for this publication using data from Alesina et al. 2020.
Note: Overall and sectoral reform indexes are continuous indicators taking a value in the (0–1) interval. A higher value indicates a greater degree of liberalization (lower intensity of restrictions). Lines represent three-year moving average fits. Structural reform is a composite index capturing the degree of liberalization in tariffs and the current account. Domestic finance reform captures the degree of liberalization in credit and interest rate controls, banking entry and supervisions, privatization, and security markets. Product market reform captures the degree of liberalization in two representative sectors: electricity and telecommunications.

difficult "last mile" improvements in policy. Nevertheless, as discussed in the following paragraphs, the latest available data across all reform areas suggest that sizeable gaps remain in reforms of middle-income and high-income EAP countries relative to a sample of developed economies (refer to figure 4.1, panels a and b, in chapter 4).

In particular, services trade restrictions have followed a declining trend across all countries. Interestingly, however, in both 2008 and 2016, several EAP countries still exhibited levels of restrictions above the ones predicted by their development level (figure 2.11). This finding comes from the World Bank–World Trade Organization Services Trade Restrictiveness Index (STRI). The STRI provides a measure of the restrictiveness of an economy's regulatory and policy framework with respect to trade in services. The STRI takes values from zero to 100, with 100 being the most restrictive regime and zero being the most open. Computation of the STRI starts from the Services Trade Policy Database, which includes a broad set of measures

FIGURE 2.11 Services trade restrictions decreased in East Asia and Pacific after 2008

Services trade restrictions, East Asia and Pacific, 2008 and 2016

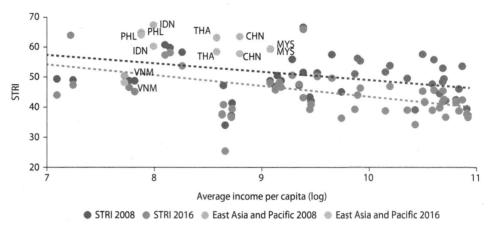

Sources: World Bank–World Trade Organization Services Trade Restrictiveness Index (STRI); World Bank, World Development Indicators; adapted from Borchert, Magdeleine et al. 2022.
Note: The figure plots for many countries the values of the STRI against the average level of GDP per capita. The blue dots and fitted line correspond to the year 2008; the orange dots and line correspond to the year 2016.

affecting services trade by both sector and mode of delivery (refer to box 2.1 for examples). About 150 of these indicators are then used in the construction of the STRI. These indicators belong to four distinct areas of policy measures: (1) conditions of market entry, (2) conditions of operation, (3) measures affecting competition, and (4) administrative procedures. The STRI is first defined at the sector-mode level by aggregating the single policy scores, and then aggregated at the sector level (refer to Borchert, Gootiiz, et. al. 2022 for details).

Although figure 2.11 provides average scores, some countries have introduced specific reforms in specific sectors. For instance, Thailand enacted several laws to reform and restructure the financial sector. These laws—including the Financial Institution Business Act (2008), the Institute of Deposit Protection Act (2008), and the Trust for Transaction in Capital Market Act (2007)—aimed to enhance the stability and efficiency of the financial sector. Viet Nam, following its World Trade Organization accession in 2007, liberalized the financial sector, opening it to foreign investors. As a third example, China's government published its first *Foreign Investment Industrial Guidance Catalogue* in 1995. Only industries included in the catalogue could receive foreign investments. Economic, technology, and environmental consulting—a service correlated to high-tech industries—was listed as a preferred industry; other services—including financial and accounting, tourism, medical services, and education—were limited or prohibited. In the 2007 updated version, services industries such as higher education service, performance art, and several financial industries were relisted as preferred industries that could receive foreign investment albeit with conditions. The 2015 update further expanded the list of preferred service industries.

Box 2.1.   What is "trade in services," and what policies affect it?

It is now customary to follow the World Trade Organization practice of taking a more comprehensive view of trade in services than that of merchandise trade. Services trade is defined to include not only traditional trade flows across borders (mode 1) but also three additional types of transactions in which supplier and consumer directly interact: the consumer moving abroad (mode 2) or the supplier moving into the territory of the consumer, either as a commercial entity (mode 3) or a natural person (mode 4). Thus, tourism, foreign direct investment, and temporary migration of service providers are also treated as modes of trade.

Accordingly, the policies affecting trade in services include measures on any of the four modes of delivery (table B2.1.1). Examples of such measures in the East Asia and Pacific region (and in other parts of the world) include restrictions on data flows that affect cross-border digital delivery of professional services, requirements to purchase mandatory insurance locally, limits on new firms or foreign ownership in telecommunications, and quotas on foreign individuals coming as independent service providers or as employees of services firms.

TABLE B2.1.1   **Examples of measures affecting services trade**

Mode of supply	Hotel services	Hospital services	Insurance services
Cross-border trade (mode 1)		Hospitals not allowed to contract foreign suppliers of telehealth services	Car and fire insurance may be provided only by domestic companies
Consumption abroad (mode 2)	Exit charge for nationals traveling as tourists abroad	No coverage of treatment abroad under public health insurance scheme	Life insurance policies purchased abroad subject to a tax of a certain percentage
Commercial presence (mode 3)	Foreign equity ownership limited to a certain percentage	Approval of new hospitals only in regions with less than a certain number of beds per a certain number of people	Number of foreign non-life insurers limited to three
Presence of natural persons (mode 4)	Number of work permits for foreign staff limited to a specific number	Doctors required to pass a test of competence with the ministry of health	No more than 10 percent of staff may be foreign nationals

Source: Original box for this publication.

An important point to consider is that services reforms that facilitate foreign direct investment can help spur the diffusion of new technologies, especially when complemented with prerequisites like high-speed broadband infrastructure (figure 2.2). Reforms and digitalization are not independent: diffusion relies on firms having the right incentives and the right prerequisites (Nicoletti, von Rueden, and Andrews 2020). Greater openness to foreign competition can spur firms to upgrade technology, and those firms with the complementary skills and high-speed broadband have the lowest costs of doing so. This result is particularly the case for data technologies, which often involve large sunk costs and depend on the high-speed collection and transmission of information.

Although these developments are interesting in themselves, key questions remain about the impact of digitalization and services reforms on productivity and jobs. The report turns to that topic in chapter 3.

Notes

1. The eight companies are Alibaba, Baidu, JD Mall, Meituan, NetEase, Pinduoduo, Tencent, and Xiaomi.
2. An important caveat regarding figure 2.9 is that it reports different definitions of access (according to the different availability of data) for different countries. Data for Malaysia, Myanmar, the Philippines, and Viet Nam are more likely to only capture access to fixed internet broadband; data for Indonesia, Mongolia, and Thailand are likely to also capture mobile internet access. This difference might explain the relatively flatness of the income gradients in panel c for Indonesia and Thailand.

References

Alesina, A. F., D. Furceri, J. D. Ostry, C. Papageorgiou, and D. P. Quinn. 2020. "Structural Reforms and Elections: Evidence from a World-Wide New Dataset." NBER Working Paper 26720, National Bureau of Economic Research, Cambridge, MA.

Borchert, I., B. Gootiiz, J. Magdeleine, J. Marchetti, A. Mattoo, E. Rubio, and E. Shannon. 2022. "Applied Services Trade Policy: A Guide to the Services Trade Policy Database and the Services Trade Restrictions Index." Policy Research Working Paper 9264, World Bank, Washington, DC.

Borchert, I., J. Magdeleine, J. Marchetti, and A. Mattoo. 2022. "The Evolution of Services Trade Policy Since the Great Recession." Policy Research Working Paper 9265, World Bank, Washington, DC.

Google, Temasek, and Bain & Company. 2022. "E-Conomy SEA Report 2022." Google, Temasek, and Bain & Company. https://services.google.com/fh/files/misc/e_conomy _sea_2022_report.pdf?utm_source=bain&utm_medium=website&utm_campaign=2022.

ITU (International Telecommunication Union). 2022. *Global Connectivity Report*. Geneva: ITU. https://www.itu.int/itu-d/reports/statistics/global-connectivity-report-2022/.

Nicoletti, G., C. von Rueden, and D. Andrews. 2020. "Digital Technology Diffusion: A Matter of Capabilities, Incentives or Both?" *European Economic Review* 128 (September): 103513.

World Bank. 2024. *Digital Progress and Trends Report 2023*. Washington, DC: World Bank.

Zhu, T. J., P. Grinsted, H. Song, and M. Velamuri. 2022. *A Spiky Digital Business Landscape: What Can Developing Countries Do?* Washington, DC: World Bank.

Why Do Digitalization and Services Reforms Matter? 3

Abstract

This chapter explores the extent to which technological change—especially in terms of adoption of digital technologies—and services reforms affect productivity, jobs, and access to education, health, and finance.

Productivity

Digitalization and productivity

The link between digitalization and productivity is as important as it is complex. Most existing economic and empirical research focused on high-income countries generally finds a positive link between digitalization and productivity.[1] New firm-level evidence from the Philippines shows that services firms that adopt digital technology increase their productivity, raise wages, and increase their value added (figure 3.1; refer to the next section titled "Jobs" for further discussion of employment results). The productivity increases are larger for the adoption of information technology or data and software capital than for the use of e-commerce. Concrete examples of these productivity-enhancing digital technologies are customer relationship management systems, data analytics adoption, and cloud computing. These digital technologies can help firms improve their productivity by streamlining processes, enhancing collaboration, enabling data-driven decision-making, and expanding market reach.

Using technology to increase productivity sometimes requires other changes, because technology adoption alone is rarely enough. For instance, adopting data technologies often requires substantial organizational change and complementary sunk investments, in both skills and infrastructure to collect and store data, and in the reorganization of processes to leverage data analytics in business decisions. A recent paper studying countries belonging to the Organisation for Economic Co-operation and Development (OECD) (Gal et al. 2019) finds that the positive

FIGURE 3.1   Technology adoption increases services firms' productivity, value added, and wages, but reduces employment

Impacts of firm technology use on firm productivity metrics, the Philippines, 2010–19

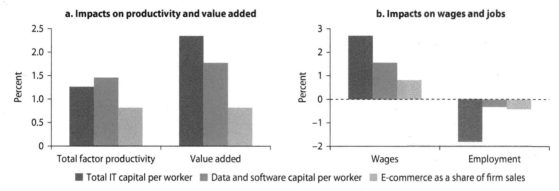

Sources: Philippines Statistical Authority: Annual Survey of Philippine Business Industry, Census of Philippine Business and Industry, and Survey on Information and Communication Technology.

Note: The figure shows results from regressions of firm performance metrics on firm technology use, including firm- and year-fixed effects. To aid comparisons across technologies, the figure presents the estimated percentage change in firm performance from a one standard deviation change in technology usage, roughly equivalent to going from the median to the 90th percentile. IT capital per worker and data and software capital per worker are expressed in logs. To avoid dropping zero values, the analysis adds one Philippine peso per worker before taking logs. Representative statistics for business sector services firms (ISIC rev 4 divisions 45–82) with 20 or more employees, for the period 2010 to 2019. All coefficients are statistically significant at the 90 percent level or more. ISIC = International Standard Industrial Classification of All Economic Activities; IT = information technology.

association between digital technology adoption and productivity is stronger in the absence of information and communication technology (ICT) skills shortages. Similarly, a recent experiment on the productivity effects of the introduction of generative artificial intelligence (AI) for customer support agents reveals a positive average productivity effect that is highest for the lowest-skilled workers, but negligible for the most skilled (Brynjolfsson, Li, and Raymond 2023).

The diffusion of digital platforms can also influence how digitalization affects productivity. Digital platforms present a competition shock for incumbent firms in their respective sectors. For example, e-commerce platforms affect traditional wholesalers and retailers by offering customers new ways of connecting with suppliers—such as through online matching, reviewing, and rating systems (Bailin Rivares et al. 2019). The Philippines has had a rapid diffusion of such platforms (figure 3.2), particularly in wholesale and retail, with major e-commerce platforms such as Grab, Lazada, and Shoppee appearing in the period 2012–15. Digital platforms in the transportation, accommodation, and travel sectors (at least before COVID-19) have had relatively more gradual diffusion.

In the case of services sectors in the Philippines, diffusion of online platforms is associated with higher productivity and with growing sales of the firms in the sector in which platforms operate (figure 3.3). Platform diffusion also affects firms that use their services, increasing those firms' productivity and raising their wages. As the impacts of platform competition ripple through supply chains, the magnitude of productivity gain is triple the size of the direct own-sector effect (refer to the next section, "Jobs," for a discussion of the employment results).

FIGURE 3.2 Online retail and transportation platforms have diffused rapidly in the Philippines

Diffusion of digital platforms, by sector, 2010–20

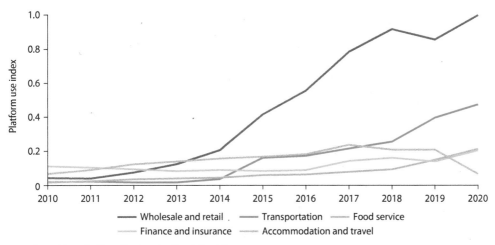

Source: Original figure for this publication using Google Trends Data.
Note: Platform use is proxied using the frequency of Google searches, following Bailin Rivares et al. (2019). The 2020 data are from the Philippine Competition Commission and are cross-checked using Philippine web traffic data from Semrush. The chart reflects 42 platforms (9 wholesale and retail, 11 transportation, 8 food service, 7 finance and insurance, and 7 accommodation and travel). Multisector platforms (for example, Grab) are allocated to individual sectors using additional keywords (for example, searches for GrabTaxi vs. GrabFood). Platform use index is normalized relative to retail platform use in 2020.

FIGURE 3.3 Online platform diffusion increases the productivity and scale of incumbent services firms in these same sectors, and leads to larger productivity gains in downstream sectors in the Philippines

Direct and indirect effects of platform diffusion on firm productivity metrics, 2010–19

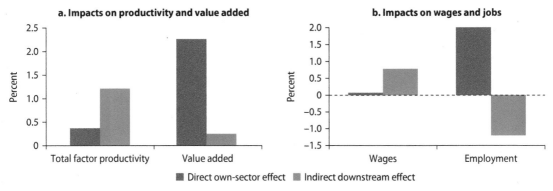

Sources: Philippines Statistical Authority: Annual Survey of Philippine Business and Industry, Census of Philippine Business and Industry, and Survey on Information and Communication Technology.
Note: The figure shows the results from regressions of firm performance metrics on measures of platform diffusion, including firm- and year-fixed effects. To aid comparisons across regressions, the figure presents the estimated percentage change in firm performance from a one standard deviation change in platform diffusion, roughly equivalent to going from the median to the 90th percentile. Direct own-sector results reflect the correlations between firm performance and platform diffusion in accommodation and travel, food service, transportation, and wholesale and retail. The downstream platform diffusion measure is a weighted sum of upstream platform diffusion, with the weights reflecting intermediate input shares taken from the 2010 OECD's Inter-Country Input-Output Tables, and excluding the use of own-sector intermediates. Indirect downstream effects are representative of manufacturing and services sectors (ISIC rev 4 divisions 10–33 and 45–82) for the period 2010–19. Wage correlations are statistically insignificant for direct own-sector effect, as are value added correlations for indirect downstream effect; all other coefficients are significant at the 90 percent level or more. ISIC = International Standard Industrial Classification of All Economic Activities.

Specifically, digital platforms can support the productivity of incumbent services firms, on average, while stimulating labor reallocation to the most productive firms. This finding is supported by the results obtained for the Philippines, which are reminiscent of the findings in Bailin Rivares et al. (2019). Those authors consider the impact of the diffusion of digital platforms on multifactor productivity of service firms in 10 OECD economies and dig deeper into the distinctions by type of platform (figure 3.4, panel a).[2]

However, the effects of digital platforms depend crucially on the type of platform considered. "Aggregator" platforms that connect existing service providers to consumers (for example, Booking.com) have tended to push up productivity, profits, and employment of existing services firms. By contrast, "disruptive" platforms that

FIGURE 3.4   In OECD countries, platforms improve productivity, but the effects depend on competition and regulation

Effects of platform diffusion on services firm productivity, by level of market concentration and regulation

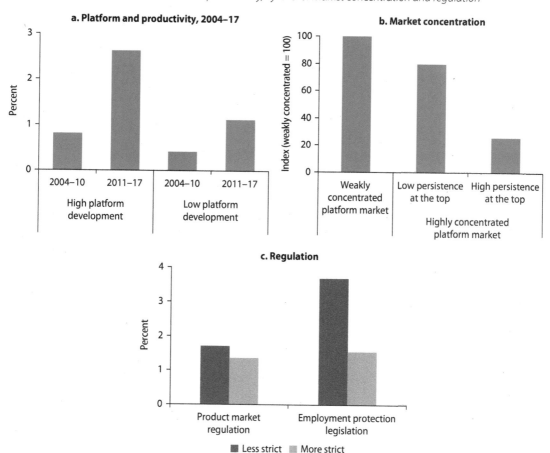

Source: Bailin Rivares et al. 2019.
Note: OECD = Organisation for Economic Co-operation and Development.

enable new types of providers to compete with existing ones (for example, Airbnb and Uber) generally have not had a significant effect on the productivity of existing providers; instead, they have tended to reduce those providers' mark-ups, employment, and wages. Finally, the overall positive effects of platforms on productivity depend on the level of competition among platforms (figure 3.4, panel b) and on prevailing product and labor market regulations (figure 3.4, panel c, where it is interesting to notice how the positive productivity effects of platform diffusion are also found in the presence of strict labor market regulations).

Overall, the existing literature and new evidence for this report suggest that digitalization can have positive impacts on productivity but that these impacts are heterogeneous and depend on complementary factors, such as skill endowments, the regulatory environment, and level of competition.[3]

Appendix F of this report includes a discussion of how digitalization is also affecting tourism, a particularly important industry in the East Asia and Pacific (EAP) region.

Services reforms and productivity

In addition to the adoption of digital technologies, structural reforms that open service sectors to greater competition can spur productivity growth in services (figure 3.5). Increased openness in services implies increased foreign presence, foreign entry, and competition between foreign and domestic providers. This competitive dynamic is expected to deliver better and more reliable provision of existing services, introduce new varieties of services, and promote competitive pricing in the services sector.

FIGURE 3.5   Services trade reforms increase productivity

Effect of services trade reforms on productivity, cross-sectional evidence from selected EAP countries and comparators, 2008–16

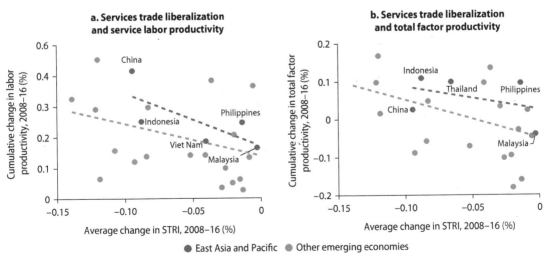

Sources: Original figure for this publication using data from World Bank, World Development Indicators; University of Groningen, Penn World Tables; and World Bank–World Trade Organization Services Trade Restrictiveness Index (STRI).
Note: EAP = East Asia and Pacific. Green dots = other emerging markets and developing economies.

These are the main channels through which services trade liberalization can lead to increased productivity in the services sector, and improved economywide performance through links between the productive sectors.

Services trade liberalization has been shown to have positive "downstream" effects on manufacturing firms (Arnold et al. 2016). Given the growing role of services in manufacturing, these positive productivity impacts from services reforms extend beyond the direct impact on services to manufacturing firms that use services. As shown in figure 3.6, analysis using data for a large sample of developed and developing countries, even after controlling for the level of development, finds that an increase in the use of services such as finance, ICT, and business services as intermediate inputs has a positive association with an increase in labor productivity in agriculture and manufacturing.

Services reform may affect the performance of manufacturing sectors in at least four ways. First, new services may become available thanks to the entry of new and more sophisticated services providers. Examples include new financial instruments and cash flow management tools, multimodal transportation services, and digital value-added services. Availability of these services may in turn lead to productivity-enhancing changes in manufacturing, such as receiving production orders online or setting up online bidding systems for suppliers. Second, services reform may increase availability of services through, for instance, extending internet coverage to rural areas. The improved access may boost performance of smaller or remotely located enterprises.

FIGURE 3.6   An increase in the use of finance, ICT, and business services is associated with higher labor productivity in manufacturing and agriculture, even after controlling for level of development

Association between use of finance, ICT, and business services and labor productivity, EAP and comparator countries, 2018

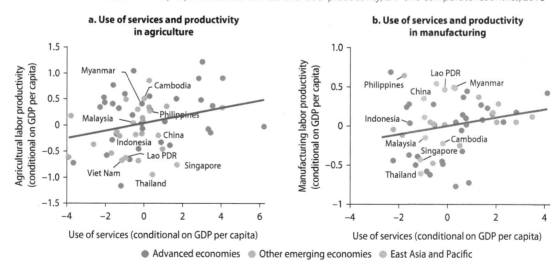

Sources: World Bank, World Development Indicators; Organisation for Economic Co-operation and Development, Inter-Country Input-Output Tables.
Note: The panels represent the positive association between the use of finance, ICT, and business services as intermediate inputs (measured as direct requirements in production) and labor productivity both in agriculture and in manufacturing. Data are for 2018. To establish that this relationship holds even after controlling for the level of development, the analysis first regressed both the use of services and the logarithm of the labor productivity on the logarithm of the GDP per capita, and then plotted in panels a and b the residuals from those regressions. EAP = East Asia and Pacific; ICT = information and communication technology.

Third, the reliability of existing services may improve because of reform. These improvements will then limit disruptions to production and lower operating costs in downstream manufacturing sectors. Fourth, reducing market power in services may enhance innovation incentives in downstream manufacturing if, before the reform, upstream service providers appropriated part of the innovation rent.

Empirical firm-level evidence from Viet Nam confirms positive productivity effects of services trade liberalization on services firms as well as positive downstream productivity effects on manufacturing firms. Since its accession to the World Trade Organization in 2007, Viet Nam has made noticeable progress to liberalize services trade. Specifically, between 2008 and 2016, the Services Trade Restrictiveness Index declined sharply in sectors such as finance, transportation, and professional services. This decline in service sector restrictiveness was associated with an average 2.9 percent annualized increase in value added per worker in each sector (equivalent to a 23.5 percent increase in labor productivity over the 2008–16 period)—figure 3.7, panel a. Furthermore, the same liberalization in services sectors is associated with a 3.1 percent increase in labor productivity of manufacturing enterprises that use services inputs, benefiting small and medium private enterprises most significantly (figure 3.7, panel b).

Analysis of firms in Indonesia and Viet Nam shows heterogenous direct and indirect productivity effects from services reforms across different firm types. The firm-level analysis from Viet Nam underlines how, in the case of services sector firms,

FIGURE 3.7   Services trade liberalization has direct and indirect productivity gains

Direct and indirect effects of services trade liberalization, by type of firm, Viet Nam, 2008–16

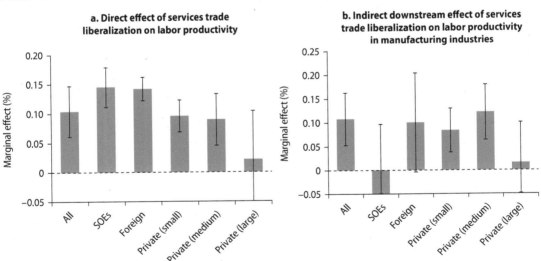

Source: Original figure for this publication using data from Viet Nam Enterprise Surveys, 2008 and 2016.
Note: Figure shows ordinary least squares regression results. The dependent variable is the change in log value added per worker between 2008 and 2016. The main explanatory variable is the change in STRI values in trade, transportation, finance, professional, and telecommunication sectors between 2008 and 2016 in panel a, and the change in the "downstream" STRI for manufacturing sectors in panel b. The downstream STRI is a sector-specific measure for each 2-digit manufacturing sector, calculated by the average STRI of the specified five services sectors weighted by the corresponding purchasing value from each manufacturing sector. The regression samples consist of all enterprises operating in trade, transportation, finance, professional, and telecommunication sectors (panel a), and all manufacturing enterprises (panel b), in 2008 and 2016. All regressions control for firms' baseline revenue and employment. Standard errors clustered at the industry level. SOEs = state-owned enterprises; STRI = Services Trade Restrictiveness Index.

the stronger effects occur within foreign firms and state-owned enterprises. Analyzing the downstream effects on manufacturing sectors suggests that small and medium private firms benefit the most from services trade liberalization. Firm-level evidence from Indonesia also shows positive but heterogeneous productivity gains in both services and manufacturing from reforms that ease regulatory constraints in services sectors. First, relaxing restrictions on specific services sectors is systematically associated with improvements in the perception of performance of those sectors. Users in downstream sectors perceived better performance by those sectors in which foreign direct investment restrictions were relaxed the most in the period after the Asian financial crisis. Second, services sector reforms contributed to increased productivity of manufacturing firms that used services intensively. A 10-percentage-point reduction in foreign direct investment restrictiveness in services is associated, on average, with an 8–9 percent increase in productivity (figure 3.8). Considering the extent of services reforms during the period, analysis reveals that these reforms added 0.4 percentage point annually to productivity growth over the period 1997–2009, or about 8 percent of the total productivity growth during the period. The gains accrued to domestic and foreign firms alike, and disproportionately more to better-performing ones. The top quartile of performers exhibited productivity increases due to services sector reforms twice as large as those experienced by median performers. Likely, better-performing firms had better capabilities and technologies to benefit from better services inputs.

To shed more light on the opportunities created by the liberalization of services trade, "out of the box" 3.1 presents the case of ICT services exports in Viet Nam.[4] This case exemplifies how some services sectors, propelled by digitalization and reform, can act as new areas of growth for developing economies and offer new opportunities of employment to young people with science, technology, engineering, and mathematics training.

FIGURE 3.8   Relaxing restrictions to FDI in services is associated with increased manufacturing productivity in Indonesia

Estimated effect of a 10-percentage-point reduction in the FDI restrictiveness index on services on TFP in downstream manufacturing

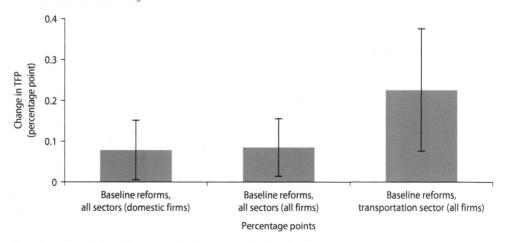

Source: Original figure for this publication using data from Duggan, Rahardja, and Varela 2013.
Note: FDI = foreign direct investment; TFP = total factor productivity.

Out of the box 3.1.   ICT services exports from Viet Nam

Almost three decades ago, Viet Nam embarked on the *doi moi* market-based reforms, not unlike those China adopted a few years earlier. After opening significantly to trade and investment, Viet Nam has emerged as one of the fastest-growing economies in East Asia. In the earlier years, commodities, fueled by Viet Nam's wealth of fertile land and water resources, spurred exports. Gradually, labor-intensive sectors like textiles and, most recently, electronics rose to the fore. Less recognized is the important role services, particularly information and communication technology (ICT) services, have begun to play in export growth and diversification.

In under two decades, the ICT sector (including software development, hardware design, information technology consulting, and other digital services) has emerged as a key driver of overall services exports growth, contributing significantly to foreign exchange earnings and employment. Viet Nam's accession to the World Trade Organization in 2007, the consequent decline in services trade restrictions, and a rise in home-grown ICT firms, saw the country move from the world's 73rd-largest exporter of ICT services in 2005 to the 38th-largest in 2021 (figure OB3.1.1, panels a and b). In the East Asia and Pacific region, Viet Nam moved from the 8th- to the 6th-largest exporter, surpassing Indonesia and Thailand.

FIGURE OB3.1.1   Viet Nam's ICT services exports have soared since 2005, and in 2021 ICT services were the third-largest contributor to overall services exports

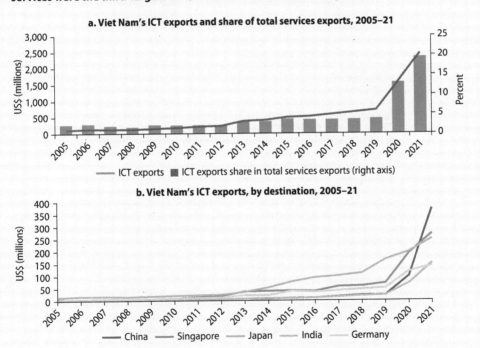

a. Viet Nam's ICT exports and share of total services exports, 2005–21

——— ICT exports   ▉ ICT exports share in total services exports (right axis)

b. Viet Nam's ICT exports, by destination, 2005–21

——— China   ——— Singapore   ——— Japan   ——— India   ——— Germany

Source: World Trade Organization, WTO Stats (https://stats.wto.org).
Note: ICT = information and communication technology.

(Continued)

Out of the box 3.1.   (Continued)

FPT Software, which began operations in 1999, is now the largest ICT services exporter in Viet Nam and an example of how technological change and economic reforms can open new areas of growth and create new employment opportunities for young people with science, technology, engineering, and mathematics training (STEM).

Excerpts from a conversation with Hoan Nguyen Khai (Senior Executive Vice President and Chief Operating Officer cum Chief Financial Officer, FPT Software)

Can you tell us a bit about the trajectory of FPT?

FPT Software (part of the FPT Corporation) started with a small team in Hanoi and has since grown to a workforce of around 30,000 people globally, operating in 30 countries and territories. Our largest overseas operations are in Japan, in multiple cities, followed by operations in the US, Canada, Europe, and the greater Asia Pacific. Out of the 30,000 employees, we have close to 25,000 in Viet Nam (graduates from top STEM and business institutions), and about 5,000–6,000 outside Viet Nam. With staff from over 20 nationalities and across multiple offices, we were one of the first companies in Asia to provide offshore, near-shore, and onshore services to our clients, as well as high-quality technical support, R&D services, and software engineering.

What would you identify as key factors in the development of FPT?

In regional markets such as Japan, several factors contributed to our development. Elements like cultural and language proximity, and time zone alignment play crucial roles. For our clients in the US, Viet Nam presents a relatively less risky environment for conducting business, especially given the ongoing geopolitical tensions with some of the larger exporting countries. Another key strength lies in the quality of our staff, a factor attributed in part to the robust STEM education system in Viet Nam mainly at the basic level. To complement this, we set up the FPT Education institution to provide continuous training and foster the supply of technology and knowledge from the young and qualified population of Viet Nam.

Can you give some specific examples of how you operate with the manufacturing customers?

We work with many manufacturers in different industries like automotive, aviation and light manufacturing. Over the years, we, and the industry as a whole, have moved from providing staffing resources for software installation and management to more advance projects involving automation and artificial intelligence. In the past, a good share of our services was delivered on client premises, but now we can provide offshore services powered by cloud-based systems. Many of our manufacturing clients need their processes migrated onto the latest systems and now with the AI trend picking up steam, we are receiving requests to support manufacturing clients to implement their AI blueprints. The most important value add when it comes to serving manufacturing clients is industry knowledge, and the ability to assemble large teams of high-quality engineers quickly.

To what extent do costs play a role in your business model (for example, competitive wages)?

In the earlier years, Viet Nam's lower wages made us attractive to our clients. However, as Viet Nam has grown so has the cost of labor. In fact, labor in Viet Nam is now more expensive

(Continued)

Out of the box 3.1.   (Continued)

compared to other countries in the region. As I mentioned earlier, we now compete on quality and speed and on scale. There are very few countries in the region that can corral a large pool of engineers in a short time.

How easy is it for you to move staff and navigate data regulations across your client countries?

It is challenging. Restrictions around staffing and data laws continue to shape our business model. For instance, our strategy of acquiring companies in key US, European, and Latin American markets is precisely because we have difficulties obtaining visas for our engineers. This is in contrast with Japan, where it is much easier for us to get work permits approved. Data sharing restrictions have also driven our strategy around satellite operations. For some of our clients, all data processing must be done on servers located with the client's country. If we had less restrictions around movement of staff and data, we would save substantial operating costs. For now, though, we have to invest in offices abroad if we want to keep building our business.

Jobs

Digitalization and reforms also affect jobs, especially in the services sector. This section presents first a deep dive into Indonesian data on digital jobs, the demand for highly skilled workers, and the associated wage premium. It then discusses the regional and global evidence on the differences between platforms for digitally delivered services and those for locally delivered services. Finally, it briefly discusses how recent developments in AI, and generative AI, may shape the future of work in the region.

In Indonesia, digital labor, defined as workers who use digital technologies and the internet for their primary work, accounted for nearly 40 percent of the workforce in 2022 and has increased rapidly as a share of digital employment in Indonesia's labor force (figure 3.9). Within digital employment, a significant increase is noticeable in the informal sector—including self-employed and individual contractors—which could be associated more directly with the rise in digital platform–based jobs.

Aggregate statistics across all sectors in Indonesia's economy also show that digital jobs, whether conducted in the formal or informal sectors, require higher levels of education than jobs not involving digital technologies (figure 3.10, panel a). Whereas close to 40 percent of formal digital workers have a university degree or above, less than 20 percent of nondigital workers do so. Across economic sectors, digital employment represents a larger share in services than in agriculture and most manufacturing sectors (figure 3.10, panel b). In addition, noticeable heterogeneity exists within the services sector. Digital employment contributes a higher share in more technical services such as ICT, financial, and insurance, and a lower share in less technical services, such as wholesale, retail, and transportation.

The skill premium for jobs involving digital technologies is likely transmitted into higher wage premiums for digital employment. Estimates from an empirical exercise that controls for various worker characteristics suggest a statistically significant

FIGURE 3.9   In Indonesia, work involving digital technologies is growing rapidly

Share of workers using digital technology for work, formal and informal sectors, 2018–22

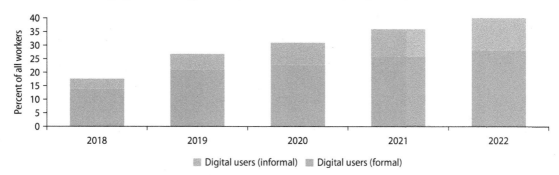

Source: Original figure for this publication based on analysis by the World Bank East Asia and Pacific Chief Economist team using Indonesia's Labor Force Surveys, 2018–22.
Note: Digital workers are defined as workers who use digital technologies and the internet for work in their primary job. Using multiple rounds of labor force surveys can offer more granular insights into how digitalization is shaping employment in the region.

FIGURE 3.10   In Indonesia, digital jobs often require a higher level of education and dominate in more technical service sectors

Digital workers, by sector and education level, 2022

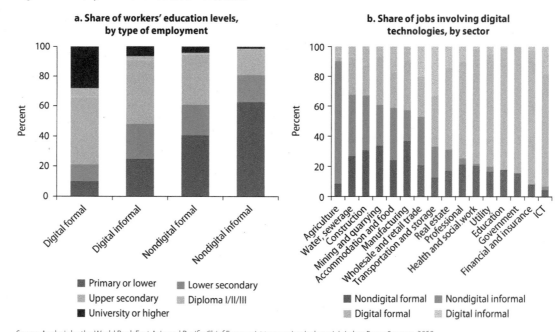

Source: Analysis by the World Bank East Asia and Pacific Chief Economist team using Indonesia's Labor Force Surveys, 2022.
Note: Digital workers are defined as workers who use digital technologies and the internet for work in their primary job. ICT = information and communication technology.

conditional wage premium of above 20 percent for jobs that require digital technologies (figure 3.11, panel a). The positive wage differential expands over the job tenure (figure 3.11, panel b), potentially suggesting the existence of learning by doing.

Looking beyond Indonesia, global evidence suggests that digitalization, specifically through digital platforms, can create new tasks and jobs but can also lead to job losses as some tasks are replaced. For example, digital platform firms have created employment for millions of people working as drivers for delivery and ride-hailing services at the low-skill end of the labor market, while boosting demand for more highly skilled labor, such as programmers and managers. At the same time, the emergence of these platforms can reduce jobs in traditional sectors.

Estimates from the International Labour Organization show a rapid increase in the number of digital labor platforms over the last decade, with the number of active platforms mediating locally delivered services (ridesharing and deliveries services) nearing 500 globally, and those mediating digitally delivered services approaching 300 (figure 3.12). The rise in digital labor platforms has provided new employment opportunities for workers worldwide. Survey estimates, which vary significantly in estimation methodologies and are currently available mostly for developed countries, suggest that the number of workers engaged in digital labor platforms either full- or part-time could have surpassed 10 percent of the population in several European Union Member States (Brancati et al. 2020) and exceeded 5 percent of the labor force in Viet Nam (figure 3.12, panel b).

Digital platform–based jobs permeate across a diverse set of service sectors and therefore vary significantly in characteristics. One way to examine the heterogeneity of these jobs is to look at differences in the level of skill intensity required.

FIGURE 3.11   In Indonesia, digital jobs earn a wage premium, which increases over time
Wages and wage premiums for digital jobs, by sector and tenure

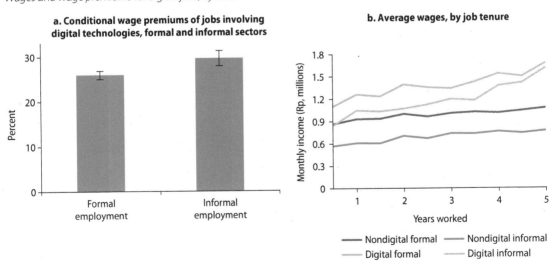

Source: Analysis by the World Bank East Asia and Pacific Chief Economist team using Indonesia's Labor Force Survey, 2019.
Note: Digital workers are defined as workers who use digital technologies and the internet for work in their primary job. Panel a shows the conditional wage premium estimated from ordinary least squares regressions controlling for individual's age group, gender, sector, location, education, and hours worked. Rp = Indonesian rupiah.

FIGURE 3.12 **Number of digital labor platforms for locally and digitally delivered services have increased significantly, and associated jobs obtained represent a nontrivial share of the workforce**

Digital labor platforms and their share of workers, globally and in selected countries

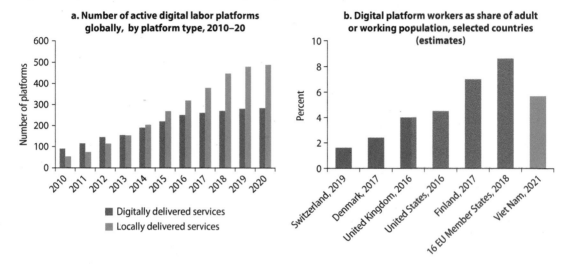

a. Number of active digital labor platforms globally, by platform type, 2010–20

b. Digital platform workers as share of adult or working population, selected countries (estimates)

■ Digitally delivered services
■ Locally delivered services

Source: Analysis by the World Bank East Asia and Pacific Chief Economist team using ILO 2021, based on Crunchbase database; World Bank 2021.
Note: In panel a, digitally delivered services include microtask, freelance, and competitive programming. Locally delivered services include ridesharing and delivery services. Statistics include only currently active platforms. Compilation in panel b based on surveys implemented with different methodologies: Brancati et al. 2020 (European Union); CIPD 2017 (United Kingdom); Farrell, Greig, and Hamoudi 2018 (United States); FSO 2020 (Switzerland); Ilsøe and Weber Madsen 2017 (Denmark); SF 2017 (Finland); and World Bank staff estimates based on Viet Nam's Labor Force Survey, 2021. Viet Nam's estimate is for location-based platform jobs. EU = European Union.

Survey statistics from a recent International Labour Organization global study show that workers in platform jobs for digitally delivered services are generally highly educated, with over 60 percent having obtained at least a bachelor's degree (figure 3.13). In contrast, jobs in locally delivered services are dominated by the less-educated workforce, with only about 20 percent having obtained higher education.

Job postings originating from EAP countries on the largest global platforms for digitally delivered services—having grown rapidly since 2017 (figure 3.14, panel a)—are concentrated in high skill–intensive service tasks such as software development and information technology or creative and multimedia, and less so in less-skilled sectors such as clerical and data entry, or sales and marketing support (figure 3.14, panel b). This distribution of skill intensity, particularly of platform-based jobs for digitally delivered service, is observable by looking at demand-side measures such as the volume of job postings across task types.

Although digital platforms have been shown empirically to create jobs, evidence also points to negative employment effects of platform diffusion. The Philippines data described in the first section of this chapter provide a preliminary perspective on this issue. As shown in figure 3.1, the direct employment effect associated with the adoption of new technologies appears negative on average. However, as shown in figure 3.3, platforms lead to an increase in employment of the firms in the sectors in which platforms operate, but employment declines in sectors that use the services of the platforms. Platforms increasingly bundle many services—for example, an e-commerce provider may offer not only retail services but also transportation

FIGURE 3.13 **Platform jobs for digitally delivered services often require higher education, but locally delivered services are dominated by a less educated workforce**

Global share of workers on digital labor platforms, by education level and platform type

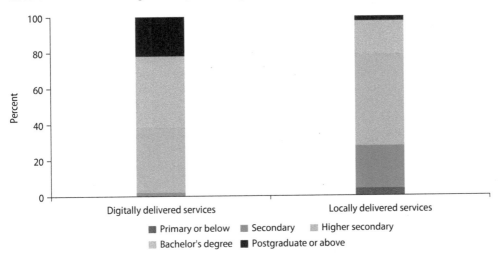

Source: Analysis by the World Bank East Asia and Pacific Chief Economist team using ILO 2021.
Note: Digitally delivered services include microtask, freelance, and competitive programming. Locally delivered services include ridesharing and delivery services.

FIGURE 3.14 **Job postings on major labor platforms for digitally delivered services have increased, especially for skill-intensive services tasks**

Share of job posting from EAP countries for digitally delivered services, by country and task type

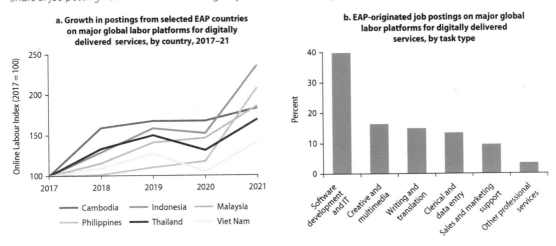

Source: Original figure for this publication using Oxford's Online Labour Index statistics for the number of job postings originating from EAP countries on the 10 largest global digital labor platforms for digitally delivered services.
Note: EAP = East Asia and Pacific; IT = information technology.

or marketing or finance or even data analytics services—with multichannel platforms like Grab offering a salient example of this bundling. The fall in downstream employment may reflect downstream firms substituting platform-provided services for services tasks that their own workers previously undertook.

Despite the importance of the question about how technologies are shaping the availability of jobs, the analysis presented in the first section of this chapter is not equipped to provide definitive answers. The effects presented in figures 3.1 and 3.3 are, in fact, average employment responses within firms to technology adoption and platform diffusion. These effects do not represent the impact on aggregate employment, which includes the effects on all the other sectors of the economy. In particular, the analysis has not addressed the impact on employment of new technologies like generative AI, which are expected both to eliminate certain tasks and to create new tasks.

The past few years have seen rapid progress in the field of generative AI and large language models (LLMs). Generative AI refers to a class of machine learning technologies that can generate new content—such as text, images, music, or video—by analyzing patterns in existing data. Some LLMs can now create content of such high quality that people cannot tell it apart from human created versions. In standard image recognition tests, LLMs had already surpassed human accuracy of 95 percent by 2016 and had accuracy of nearly 99 percent as of 2021 (Zhang et al. 2021). To date, generative AI models have mostly been deployed to perform specific tasks such as generating images from captions or transcribing text from speech. However, LLMs can create additional tools and are increasingly being integrated into specialized applications in fields as broad as writing assistance, coding, and legal research.

This new wave of AI has already had a wide range of impacts on labor markets in East Asia and Pacific (EAP) countries and is expected to have more. Box 3.1 explores the channels through which AI could affect labor markets in the EAP region.

The impacts of AI in EAP countries will depend heavily on the policy choices made to facilitate and steer its deployment. Effects will depend first on rates of AI adoption. According to evidence from the Stanford AI Index, adoption of AI in most EAP countries, except in the Republic of Korea and in Singapore, remains far lower than in advanced economies, which suggests that AI will have limited impact in EAP countries without further efforts to stimulate diffusion of AI and without providing foundational skills and training for workers to benefit from AI.

The constructive and destructive applications of AI are boundless, and EAP countries will need to carefully formulate policies to encourage the use of AI to benefit public services, generate good jobs, and stimulate research in socially beneficial areas. Many EAP countries also face demographic challenges with significant population aging, meaning they could run out of workers before they run out of jobs. With the right policy choices, generative AI has the potential to offer solutions to some of these challenges, and many others, for example through increasing the quality of health care and education.

In conclusion, although the effects of new technologies on the total number of jobs remain uncertain, this section has highlighted how the demand for skills will likely increase with the adoption of new technologies. "Out of the box" 3.2 presents the case of offshore services in the Philippines. Conversations with several services firms there confirmed how the diffusion of new technologies—among their clients and within the firms themselves—has increased the need for staff equipped with more sophisticated skills, including creative thinking.

Box 3.1. How might artificial intelligence affect labor markets in East Asia and Pacific?

Task displacement. For some occupations, artificial intelligence (AI) will displace certain tasks and, in some cases, may displace the full spectrum of tasks. Felten, Raj, and Seamans (2023) constructed a leading measure of this potential for displacement at the occupation level, with exposure to AI defined as the degree of overlap between AI applications and required human abilities. They show that, whereas previous waves of automation disproportionately affected jobs involving routine tasks (data entry, bookkeeping, assembly line work, and others), generative AI can perform a variety of nonroutine tasks (coding, persuasive writing, graphic design, and others). No clear relationship exists between the types of jobs exposed to generative AI and the types of occupations exposed to earlier waves of automation (figure B3.1.1, panel a).

Many EAP countries have low workforce exposure to displacement from generative AI. Pizzinelli et al. (2023) extend the occupation-level exposure measures of Felten, Raj, and Seamans (2023) to measure exposure at the country level. Among the nine EAP economies included in their research, the median share of the workforce in low exposure occupations is 66 percent, compared to only 40 percent in advanced economies (figure B3.1.1, panel b). This finding reflects the high share of the workforce in EAP economies in sectors with low generative AI exposure, such as agriculture and construction.

FIGURE B3.1.1   Exposure to generative AI is different from exposure to previous waves of automation, and many EAP countries have low exposure

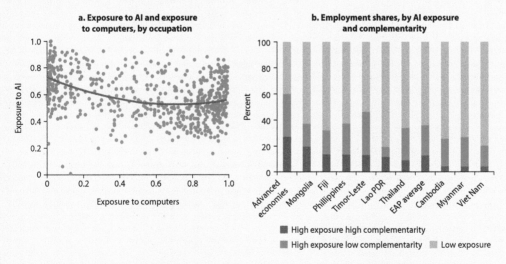

Sources: Cazzaniga et al. 2024; Felten, Raj, and Seamans 2023 (exposure to AI); Frey and Osborne 2017 (exposure to computers).
Note: Each dot in panel a represents an occupation (defined according to US SOC 2010), with exposure to AI rescaled to range from 0 to 1. Panel b uses measures constructed by Cazzaniga et al. (2024) using the AI exposure scores of Felten, Raj, and Seamans (2023) and the complementarity index of Pizzianelli et al. (2023). AI = artificial intelligence; EAP = East Asia and Pacific.

(Continued)

Box 3.1.   (Continued)

Task creation. New technologies also generate new tasks that increase labor demand. Roughly 60 percent of US employment is found in occupations that did not even exist in 1940 (Autor 2024). Similarly, AI can create new varieties or higher-quality products that require new types of worker tasks to produce. The most obvious example is the demand for labor related to building, training, and applying large language models. In the longer term, however, the new tasks created from generative AI could be wide-ranging.

Labor productivity augmentation. AI will also augment human performance in certain tasks needed to work with AI, increasing their productivity rather than replacing them. Pizzinelli et al. (2023) recently attempted to measure augmentation with an index of potential AI complementarity that reflects an occupation's likely degree of shielding from AI-driven job displacement. For example, because of advances in textual analysis, judges are highly exposed to AI but are also highly shielded from displacement because society is unlikely to delegate judicial rulings to unsupervised AI. Consequently, AI will likely complement judges, increasing their productivity rather than replacing them.

Along with a low proportion of the workforce in occupations highly exposed to AI, the EAP countries studied also have a low proportion in occupations likely to be complemented by AI. In these countries, the median share of the workforce in high-exposure, high-complementarity occupations is only 13 percent, compared to 27 percent in advanced economies (figure B3.1.1, panel 1b)—reflecting the low share of the workforce in these EAP countries in cognitive jobs with a high degree of responsibility and interpersonal interactions, such as those performed by surgeons, lawyers, and judges.

AI could also augment less-skilled workers in EAP countries by automating skills they don't possess and allowing them to do new types of work. AI has already semiautomated certain tasks: people without coding expertise can code with AI-assisted copilots, and people with limited English can draft fluent text with generative AI. Consequently, AI could help developing countries overcome certain skills bottlenecks. Less-skilled workers who can use AI may have a comparative advantage in jobs previously open only to more highly skilled workers, thus increasing wages of less-skilled workers. Several papers find the largest AI-related productivity gains for the least skilled and least experienced workers, in contexts ranging from customer services, professional writing, and taxi driving (Brynjolfsson, Li, and Raymond 2023; Kanazawa et al. 2022; Noy and Zhang 2023). For this reason, Autor (2024) argues that, by democratizing niche expertise that commands a high wage premium, AI could benefit middle-skill, middle-class workers.

Wider impacts. As a "general purpose technology," generative AI could have a range of impacts on innovation, technological progress, and productivity growth through diverse areas including drug discovery and increased research efficiency. Evidence suggests that the pace of innovation has slowed in recent years, thus the potential impact of AI on innovation efficiency is of first order importance for long-term growth in EAP economies (Trammell and Korinek 2023).

AI could affect labor markets in EAP countries through several other channels: income effects that increase trade, increased consumer demand for particular goods, and reallocation of employment across occupations, sectors, and countries. Continued improvements in machine translation from generative AI could further increase cross-border data flows and services trade, increasing outsourcing and facilitating greater cross-country collaboration (for example, Brynjolfsson, Hui, and Liu 2019).

Out of the box 3.2.   Offshore services in the Philippines

Business process outsourcing (BPO) helped lead the Philippines' shift from an agriculture-focused economy to a service-oriented one. In under two decades, the sector expanded to over 1.5 million employees and has served as one of the country's largest foreign exchange earners (figure OB3.2.1, panels a and b). In the early years, call centers played a pivotal role, supported by a large pool of cost-effective English-speaking workers, low telecommunication costs, and legal and educational system similarities with the United States.

FIGURE OB3.2.1   The BPO sector has been a major driver of services exports and growth in the Philippines

Service exports in the Philippines, by value, share of GDP, and composition

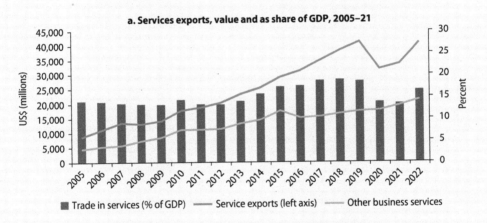

a. Services exports, value and as share of GDP, 2005–21

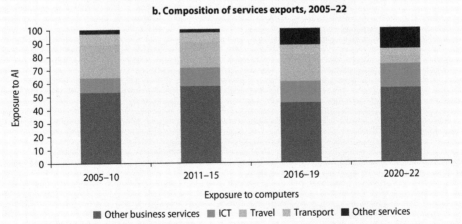

b. Composition of services exports, 2005–22

Source: World Trade Organization, WTO Stats (https://stats.wto.org).
Note: "Other business services" encompasses business process outsourcing (BPO) services. ICT = information and communication technology.

(Continued)

Out of the box 3.2.   (Continued)

As the traditional BPO sector matured, firms looking to boost their competitiveness diversified into higher-value workstreams such as nonvoice processes (email, chat, and back-office operations). Business process management has also taken off, with firms managing more complex activities in finance, legal, and human resources (HR). This evolution, together with a rapidly changing technological landscape, seems to have increased the sector's demand for higher-skilled labor.

Penbrothers, Sprout Solutions, Tata Consultancy Services (TCS), and Thinking Machines Data Science occupy different segments of the offshore services value chain, but all four rely on technology for service delivery. Penbrothers, a full-service employer of record, helps clients build and manage remote offshore teams. Sprout Solutions offers HR and payroll management services geared toward the Philippines complicated HR compliance requirements. TCS, a multinational BPO firm, supports some of the world's largest companies. Thinking Machines Data Science provides technical services like geospatial analysis, data science, and machine learning solutions.

What types of skills do you look for when recruiting? How is this evolving with changes in technology?

Carlo Lim (Director, Penbrothers): Typically, we recruit high-skilled workers with at least a bachelor's degree and an average of five years of relevant on-the-job experience. However, most of our employees serve nonmanagerial roles in finance, HR administration, customer resource management, technology, and other general support roles. We hire on a bespoke basis for our clients and offer competitive salary packages and benefits. On the impact of technology, we have noticed, with the rise of artificial intelligence (AI), more requests for training on AI tools and a higher interest in hiring talent that is already familiar with industry specific trends around AI.

Patricia Lim (Vice President, Operations, Thinking Machines Data Science): Since our inception we have focused on hiring talent from a slew of academic and professional backgrounds with an emphasis on strong critical thinking skills. Over the years our operating model has allowed more junior staff robust opportunities to learn by doing and through working closely with more experienced technical and management leads. However, the pipeline for talent is narrow especially when it comes to soft skills. And as technology becomes more sophisticated, critical thinking and creativity will matter even more in how we serve our clients. If the Philippines aims to produce even stronger talent, education policies will need to emphasize both math and the arts and not just at the basic level but all the way through college.

What role has technology played in your business model? How is AI reshaping this? And what kind of talent do you need to continue to thrive?

Kislay Chandra (Chief Operations Officer, Sprout Solutions): Technology and talent are really at the core of how we have managed to create an HR and payroll software solution that helps our clients in the Philippines navigate the country's complex HR regulatory requirements. A key inflection point for our business came when, after a few years of managing our own data servers, we decided to switch over to cloud computing. By outsourcing our data storage

(Continued)

Out of the box 3.2.   (Continued)

and computing needs, we were able to cut down on costs and, at the same time, sustainably scale up our business. Now we are following developments around AI very closely and have started including AI-powered search functions in our software. This could possibly be another critical moment in our journey, but we are also acutely aware of the downside risks. If we want to seize the opportunities around AI, we will need staff that are tech savvy and strong critical thinkers. We have managed to build our company with mostly Filipino talent, however as the technology gets more complex, the talent pool gets smaller, and it has been difficult to compete with foreign companies for the best engineers and managers.

Have the technology needs of your clients evolved over the years, and how has TCS Philippines responded to client demands?

Shiju Varghese (TCS Philippines Country Head): Over time, client needs have evolved and more so recently with the rise of AI. Customers are increasingly asking for AI-centered solutions to help streamline business processes and cut costs alongside requests for cutting-edge cloud and cybersecurity solutions. To stay competitive, TCS has had to adapt, and we have rolled out training, at varying levels of technicality, on generative AI to upskill and prepare staff to respond to customer needs. In terms of incorporating technology into our work, we have benefitted from being part of a conglomerate with delivery centers across the world. It has enabled us to provide our teams and our customers with a sandbox to pilot new products and ideas, and we also have the space to scale up innovative solutions quickly across different contexts. Cloud technologies and capabilities have been central to this global delivery model and will continue to be.

Access

Digitalization and reforms matter not only for jobs and productivity but also for access to key services, such as education, health, and finance. This section explores how the interplay between technology and policy can provide better access to these services. These services play a pivotal role in augmenting the human capital (capacities) needed to harness the greater opportunities arising from higher productivity and demand for skills—opportunities explored in the previous sections.

Education

Investments in literacy and numeracy skills have been pivotal in spurring economy-wide productivity growth, serving as a cornerstone for EAP's economic miracle (World Bank, forthcoming). However, future growth in middle-income EAP countries is at risk given persistent low levels of learning. According to 2018 data for eight EAP countries (Cambodia, Kiribati, the Lao People's Democratic Republic, Myanmar, Papua New Guinea, the Philippines, Tonga, and Tuvalu), more than 66 percent of children struggled to read or understand age-appropriate texts. Notably, even in upper-middle-income countries like Malaysia, this figure exceeds 40 percent (figure 3.15, panel a).

FIGURE 3.15   Significant gaps in foundational learning outcomes exist between high- and middle-income countries in the region and among students of varying socioeconomic status
Learning poverty rates and learning outcomes, EAP countries, 2018

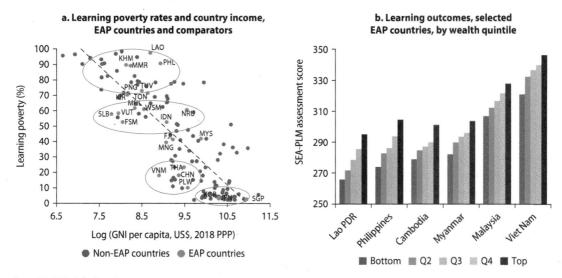

a. Learning poverty rates and country income, EAP countries and comparators

b. Learning outcomes, selected EAP countries, by wealth quintile

Source: World Bank, forthcoming.
Note: Panel b shows average proficiency score in Southeast Asia Primary Learning Metrics (SEA-PLM) points by national wealth quintiles. EAP = East Asia and Pacific; GNI = gross national income; PPP = purchasing power parity.

Conversely, high-income countries in the region such as Japan, the Republic of Korea, and Singapore all report impressively low learning poverty rates, ranging between 3 percent and 4 percent. Even within countries, students from disadvantaged socioeconomic backgrounds demonstrated poorer foundational learning outcomes than their wealthier peers (figure 3.15, panel b).

The growth of technological infrastructure and innovation holds the potential to help education systems improve student learning outcomes and help mitigate learning inequality. Particularly helpful in making economic systems more productive, technology also has the potential to help education systems produce more learning. However, it can do so only if education systems both (1) align with student learning and (2) have the capacity to use the technology effectively, which require both education system capacity and system policy alignment with learning.

Unfortunately, widespread adoption of effective education technology (edtech) in the school systems of middle-income EAP countries remains in its early stages (Yarrow et al., forthcoming). Existing evaluations (some described in box 3.2) highlight the variable impacts of edtech on student learning outcomes in these countries, from negative (−0.07 standard deviation of outcomes) to positive and large (0.98 standard deviation)—figure 3.16. Many of the available evaluations test the efficacy of treatments on a small scale, rather than the effectiveness of edtech programs implemented by ministries of education on a large scale. As illustrated in the variety of impacts, a mix of factors beyond the efficacy of a specific program (such as implementation capacity, teacher compliance, and usage intensity) led to significantly different impacts.

FIGURE 3.16   Existing studies from middle-income EAP countries show large variation in effect sizes of education technology programs on student learning

Size of effect of education technology programs on student learning, by type of treatment

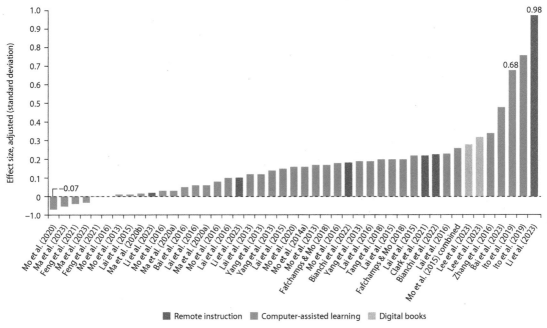

Source: Yarrow et al., annex 1, forthcoming. See annex 1 for a complete list of the studies and effects sizes (adjusted).

Edtech can be effectively incorporated into the three key policy levers (strengthen teacher selection, enhance teacher capacity, and encourage greater teacher effort) identified in the World Bank's *Bridging the Basic Learning Gap* flagship report (World Bank, forthcoming) to best enhance student learning outcomes. Recruiting teachers who are committed to improving the learning outcomes of all students using a range of tools including tech can help improve system performance. Teacher capacity requires urgent attention because most of the teachers expected to be employed in the region by 2030 are already in the system (World Bank, forthcoming). Empirical evidence from the region indicates that edtech can support improvements in teaching practice (for example, Chen et al. 2020) and be combined with nontech interventions, such as financial incentives and teacher performance scorecards, to foster teacher engagement with communities and elevate student learning outcomes. To maximize the long-term benefits of edtech, policy makers will need to support the development, implementation, and evaluation of edtech programs at scale, and align adoption and implementation with learning goals.

David (1990) points to the delayed impact of technology on economic productivity: computers and electricity, among other inventions, required the creation of networks and systems before they could observably affect productivity statistics. Similarly, edtech holds significant potential for enhancing learning outcomes and,

Box 3.2.   Highlights of studies on the impacts of education technology

Policy makers can learn from studies to better use education technology (edtech) interventions to improve public education service provisions. Of the following studies, the first demonstrates the importance of ensuring that edtech programs are properly implemented and monitored, and do not crowd out other learning activities. The second example shows the need for adequate infrastructure and improved teacher training, and the third offers an example of an edtech program that leverages top-performing teachers to bolster the teaching quality of their low-performing counterparts and enhance academic outcomes for rural students.

- The government-led computer-assisted learning program in rural China, examined by Mo et al. (2020), failed to improve student achievement or teacher training quality. Compared to schools that employed the same treatment implemented by a nongovernmental organization, student scores decreased in the 40 schools with government-implemented treatment (−0.07 standard deviation) after one academic year. Teachers in the government program tended to substitute the edtech intervention for regular instruction and did not follow the required protocol, likely because government officials were less inclined to directly monitor the implementation process. In this context, institutional capacity and commitment played a determinant role, independent of the tech intervention itself.

- Work with a computer-based innovative math learning program in Cambodia showed large impacts on student learning but was not implementable at scale (Ito et al. 2019). In five rural schools with 1,600 students, the evaluation found an impact of 0.68 standard deviation increase in student math scores. However, program implementation required an additional person to provide additional in-class support and drive the computers to and from school each day, because the schools had no electricity to charge them. This example shows the importance of physical infrastructure (electricity) as well as teaching skills to effectively implement edtech approaches, regardless of the merits of the approaches themselves.

- An evaluation of a dual-teacher program in China shows the potential of computer-assisted instruction to improve student academic outcomes and, crucially, teacher instructional practices in poor and remote areas (Li et al. 2023). Lecture videos and teaching materials from urban elite schools were made available through the internet to more than 200 remote Chinese middle schools. Teachers in nine schools included in the evaluation spent an average of 5.25–7.25 hours per week watching these lecture videos for their class preparation, which allowed them to improve their own teaching practices over the long term. In addition, a subset of students were shown lectures by expert teachers and exposed to improved teaching practices. The difference-in-differences analysis found substantial improvement in learning outcomes—a 0.98 standard deviation increase in student math scores over the three-year middle school education—a product of both the direct and indirect effects of computer-assisted instruction.

consequently, future economic growth, though widespread adoption of effective edtech in the school systems of middle-income EAP countries is still in its early stages (Yarrow et al., forthcoming).

Health

Access to good quality health care provides the foundation not only for individual well-being but also for countries' sustainable economic growth. When children are healthier, they are happier and learn better, and are more likely to become productive workers, live a long healthy life, and invest in the health and education of their own children. Investing in population health lays a solid foundation for sustained economic growth across generations (Bloom and Canning 2008).

Chronic diseases such as diabetes, hypertension, and heart-related illnesses have emerged as the main causes of ill health in EAP, partly because of aging and changing lifestyles. In developing EAP countries, the impact of chronic diseases when measured as a share of total years lost to illnesses and early deaths (disability-adjusted life years) increased from 54 percent to 80 percent over the period 1990–2019. In 2019, 85 percent of total disability-adjusted life years were related to chronic diseases in China; 72–77 percent in Fiji, Indonesia, Malaysia, Thailand, and Viet Nam; and 60–65 percent in Cambodia, Myanmar, and the Philippines (IHME 2023). As societies have aged, the incidence of chronic diseases has risen (Barnett et al. 2012). Lifestyles have changed in favor of chronic diseases, partly as high-caloric foods become easily available and sedentary lifestyles prevail. To cope with rising chronic diseases, health care systems need to move away from disease-based models of care and toward primary health care with a focus on person-centered integrated systems. Primary health care can prevent and detect diseases early and provide continuous personalized care at affordable costs for a wide share of the population. Reinforcing the role of primary care physicians as gatekeepers or the primary source of care could lower the growth of health care expenditures and enhance overall efficiency of health care systems (OECD 2020; WHO 2018).

Health information technology has the potential to accelerate reforms toward integrated health care systems, especially through reforms that make data sharing easy. Digital technology facilitates the quick creation, management, and analysis of electronic medical records for lower costs compared to traditional labor-intensive approaches. Secured electronic medical records can be shared and accessed among health service providers through health information exchange systems that help health professionals from primary to tertiary levels collaborate more effectively, ultimately leading to better health outcomes (refer to box 3.3 for a case study from China).

Health information technology could also incentivize health service providers to be more person-centered, which would encourage patients to seek and continue care. Electronic medical record systems free health professionals from time-consuming paper-based processes that require space for storage and make document sharing difficult. For instance, combining telemedicine with electronic medical record systems would help health professionals spend less time on administrative tasks and

Box 3.3.   Integrated care intervention in Henan, China

In Henan province, China, an integrated care intervention using health information technology as a main component has enhanced the accessibility and continuity of health services and improved the chronic conditions of patients.

Issues. More and more people bypass primary health facilities and seek care in hospitals, which leads to inefficiency by diverting limited hospital resources to mild cases that primary care facilities could manage. The limited contact between patients and health professionals in hospitals weakens the continuity of care, important for effective long-run management of chronic conditions. Bypassing primary health facilities for hospitals also increases costs for both patients and health care systems.

Intervention. In a system that integrates health facilities at primary, secondary, and tertiary levels, physicians refer patients to other levels according to the severity of the cases (figure B3.3.1). Health facilities share patients' electronic medical records using a health information exchange system. Chronic cases are managed using computerized clinical pathways. Financing incentives are implemented. For care providers, performance-based payment is used. For patients, co-payment is significantly lowered for care in primary care facilities, with a discount applied if patients are referred from primary care facilities.

Impacts. The condition of patients with diabetes or hypertension in the intervention group significantly improved as compared to those in the control group (figure B3.3.2). The share of patients who expressed concerns about costs significantly decreased, and patients' satisfaction with travel time improved. A higher share of patients responded that health professionals encouraged them to ask questions, which promotes the continuity of care. Cooperation via referrals across different tiers of health facilities improved significantly.

FIGURE B3.3.1   Integrated care delivery in Henan province, China, 2008–14

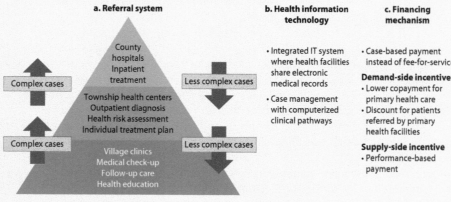

Source: Original figure for this publication.
Note: IT = information technology.

(Continued)

Box 3.3. (Continued)

FIGURE B3.3.2 **Impact of integrated care delivery in Henan province, China**

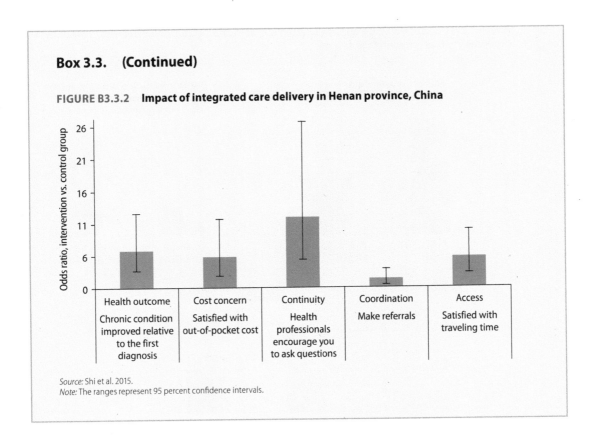

Source: Shi et al. 2015.
Note: The ranges represent 95 percent confidence intervals.

more time on communicating with patients over the screen. On the patient's side, telemedicine would lower barriers for health care access by reducing the time and money patients would otherwise spend on in-person doctor visits. This convenience could encourage people to seek care early and motivate patients with chronic conditions to maintain longer-term follow-up care regimes.

To promote more efficient health care systems, EAP countries need to accelerate the implementation of health information technology. The World Bank survey in 2023 shows that developing EAP countries have started to adopt health information technology and plan to adopt and expand it in the next five years. Currently, electronic medical record systems are implemented more widely for secondary and tertiary health facilities than for primary health facilities (figure 3.17, panel a). Few countries have adopted health information exchange systems that allow data sharing among health facilities and related organizations, but most countries have plans to do so eventually (figure 3.17, panel b). Telemedicine is available more in secondary and tertiary health facilities than in primary health facilities (figure 3.17, panel c). Table 3.1 presents technology adoption, and plans for adoption, by type of health facility and by country.

FIGURE 3.17 Secondary and tertiary health facilities use electronic medical record systems more than primary health facilities do

Status of health information technology in EAP countries, by type of facility, 2023

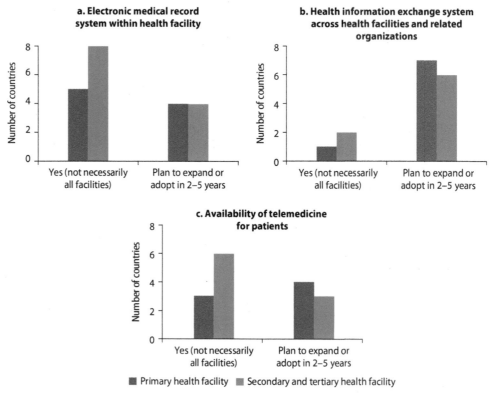

Source: World Bank survey, 2023.
Note: Eight countries participated in the survey: Cambodia, China, Indonesia, the Lao People's Democratic Republic, Mongolia, Thailand, Tonga, and Viet Nam. EAP = East Asia and Pacific.

Successful adoption and implementation of digital technology requires an overarching policy framework, sound governance, and sustained investment. Unlike other sectors such as finance and media, the health sector lags in adopting technologies to meet the public's changing needs for integrated, person-centered, and affordable care. The long-standing fragmentation of health care systems, one of the main barriers to technology adoption, causes reluctance and difficulty in communicating across different parts of the system. To address this barrier and others, policy makers will need to develop a holistic policy framework for implementing digital technology in building integrated health care systems, ensure effective governance to implement this policy framework, and encourage sustained investment.

TABLE 3.1 Status of health information technology in EAP countries, by type of facility, 2023

Type of facility	Electronic medical record system within health facility		Health information exchange system across health facilities and related organizations		Availability of telemedicine for patients	
	Yes (not necessarily all facilities)	Plan to expand or adopt in 2–5 years	Yes (not necessarily all facilities)	Plan to expand or adopt in 2–5 years	Yes (not necessarily all facilities)	Plan to expand or adopt in 2–5 years
Primary health facility	China, Indonesia, Thailand, Tonga, Viet Nam	Cambodia, Indonesia, Lao PDR, Mongolia	Tonga	Cambodia, China, Indonesia, Lao PDR, Mongolia, Thailand, Viet Nam	Cambodia, Indonesia, Thailand	China, Mongolia, Tonga, Viet Nam
Secondary and tertiary health facility	Cambodia, China, Indonesia, Lao PDR, Mongolia, Thailand, Tonga, Viet Nam	Cambodia, Indonesia, Lao PDR, Mongolia	Thailand, Tonga	Cambodia, China, Indonesia, Lao PDR, Mongolia, Viet Nam	Cambodia, China, Indonesia, Mongolia, Thailand, Viet Nam	Lao PDR, Mongolia , Tonga

Source: World Bank survey, 2023.
Note: Eight countries participated in the survey: Cambodia, China, Indonesia, the Lao People's Democratic Republic, Mongolia, Thailand, Tonga, and Viet Nam. EAP = East Asia and Pacific.

Finance

Digital technologies are also rapidly transforming the financial sector landscape. Alongside Central Asia, Europe, and Sub-Saharan Africa, EAP is one of the most dynamic regions when it comes to financial technology (fintech). The use of digital payments—the most basic type of fintech service—has expanded rapidly over the past decade. In 2021, nearly 80 percent of the adult population had reportedly received a digital payment, although with large variation across countries (figure 3.18). To put this growth into perspective, the EAP region accounted for half the world's global digital payments revenue in 2021 and for 57 percent of global revenue growth, although 88 percent of EAP's growth is concentrated in China.

The region has also had high fintech investments: in 2021, digital financial services took over as the top investment sector in Indonesia, Malaysia, the Philippines, Singapore, Thailand, and Viet Nam (Google, Temasek, and Bain & Company 2022). Policy makers and regulators in the region have supported the expansion of digital financial services through various initiatives and regulatory sandboxes. Not all EAP countries exhibit high levels of fintech activity, however. Whereas China stands out for its high fintech activity, the Lao People's Democratic Republic and Myanmar

FIGURE 3.18 **The use of digital payments has expanded rapidly since 2014**

Adults who made or received a digital payment in the past year, EAP countries, 2014, 2017, and 2021

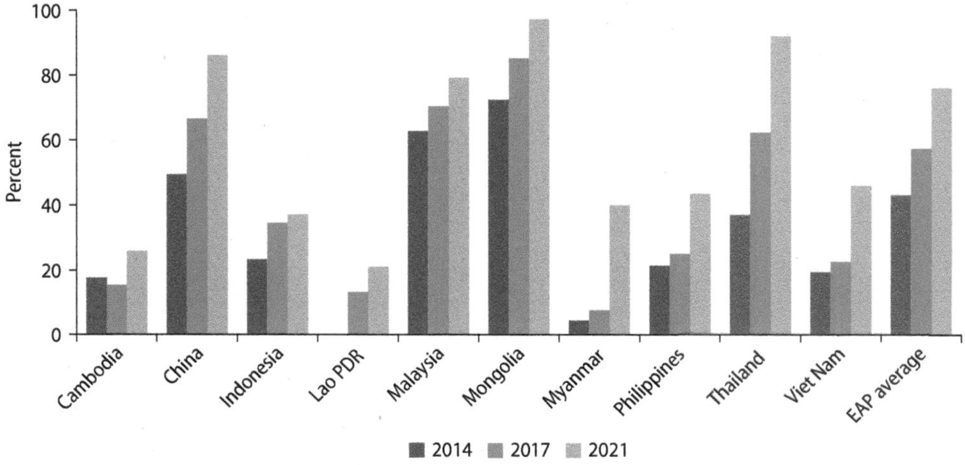

Source: Demirgüç-Kunt et al. 2022.
Note: Data for adults ages 15+. EAP = East Asia and Pacific.

appear in the bottom-five developing countries, along with Ethiopia, Morocco, and Pakistan. Policy makers and regulators in the region have supported the expansion of digital financial services through various initiatives and regulatory sandboxes (box 3.4).

Digitally enabled financial services and business models are reshaping consumption patterns, enabling wider access and adoption, including among poorer consumers. For example, some 340.7 million adults, or 20 percent of the adult population (ages 15+), in EAP do not have a traditional bank account. Reflecting costs and other barriers in traditional banking services, access to traditional bank accounts is strongly correlated with per capita income across countries (figure 3.19). In contrast, the adoption rate for mobile money accounts does not seem to vary with per capita income levels across countries, enabling consumers in lower-income economies to leapfrog.

The diffusion of fintech is uneven. For example, the use of digital payments—the most widely available digital financial service—is not uniform across the region. According to Worldpay (2022), cash is still the top in-person point-of-sale payment method in several countries including Thailand (where it accounts for 63 percent of point-of-sale transaction value), Viet Nam (54 percent), Indonesia (51 percent), and the Philippines (48 percent). Although some EAP economies have achieved high levels of digitalization around private sector wage payments, others still have room for growth. In China and Thailand, two upper-middle-income economies, about 45 percent of adults received a private sector wage payment in 2021, and the vast majority (about 80 percent) received it into an account. In Cambodia, Indonesia, Lao PDR, and the Philippines, all lower-middle-income economies, about 25 percent

Box 3.4. Innovation facilitators in East Asia and Pacific

Policy makers in the East Asia and Pacific (EAP) region have been using "test and learn" approaches to encourage financial technology (fintech) innovation through various innovation facilitators. The following examples show approaches adopted by EAP policy makers to bring new firms and business models within the regulatory perimeter in a responsible and controlled manner.

In Malaysia, the Bank Negara Malaysia started supporting financial innovation by establishing a Financial Technology Enabler Group in June 2016. This group then launched a fintech regulatory sandbox in October 2016 that permits the testing of innovative products, services, and business models.

In Indonesia, both the central bank, Bank of Indonesia (since 2017), and the financial services authority, the OJK (since 2018), offer regulatory sandboxes for firms under their remit. The Bank of Indonesia sandbox focuses on "forward looking" fintech services whereas the OJK sandbox focuses on fintech firms helping support financial inclusion and literacy. The Bank of Indonesia has had a dedicated fintech office since November 2016.

Other EAP jurisdictions like China, the Philippines, and Thailand are using regulatory sandboxes. The Pacific Regional Regulatory Sandbox was launched in 2020 for cross-border testing of fintech solutions. Members include Fiji, Papua New Guinea, Samoa, the Seychelles, the Solomon Islands, Timor-Leste, Tonga, and Vanuatu.

FIGURE 3.19   Leapfrogging in the adoption of digital finance: Access to traditional bank accounts depends on the level of income, but access to mobile money accounts does not

Access to traditional bank accounts and mobile money accounts, by income level, selected economies, 2021

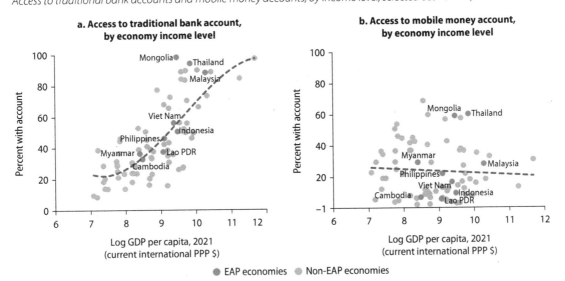

Source: World Bank, Global Findex Database and World Development Indicators.

of adults received a private sector wage payment, with the share of wage earners receiving the payment into an account ranging between 17 percent in Lao PDR and 41 percent in the Philippines (figure 3.20).

Moreover, as digital payments have become ubiquitous in many countries, other financial services have also expanded but still have lower penetration. Growth in fintech lending lags digital payments but is picking up in a few countries. China, the United States, Europe, and Australia have been leading in fintech activity worldwide, but EAP, Latin America and the Caribbean, and Sub-Saharan Africa are catching up (map 3.1). Countries in EAP have historically had higher activity levels for finance app downloads and have displayed high levels of both absolute and relative financial app downloads. The region—especially more developed economies like Australia; Hong Kong SAR, China; and Korea—has also seen a rise in person-to-person lending marketplaces.

Technology companies—as opposed to traditional financial service providers, such as banks and insurance companies—are playing an increasingly prominent role as providers of digital financial services. Big platform firms have disrupted traditional financial service delivery, leveraging their superior technological capabilities and large existing customer base to achieve scale rapidly. Particularly in some parts of EAP—such as China (Alibaba, Baidu, and Tencent) and Indonesia (GO-JEK)—these big platform firms offer a suite of financial services. Some jurisdictions in EAP (such as China, Korea, and Singapore) have also allowed platform firms to hold shares in digital banks. Credit provision by such nontraditional arrangements has accelerated in countries like China and Indonesia, bringing a new set of challenges

FIGURE 3.20 **Use of digital private sector wage payments varies widely across the EAP region**

Private sector wage payments, EAP region and upper-middle-income economies, 2021

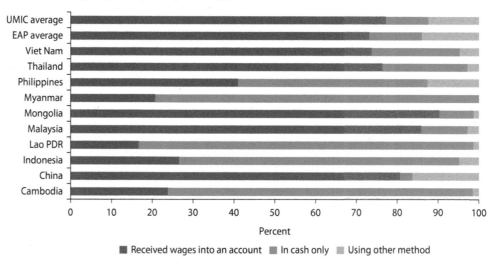

Source: World Bank 2022.
Note: The figure shows the mode of receipt of private sector wages by adults (ages 15+) in 2021 (expressed as a percent of adults who received private sector wages). EAP = East Asia and Pacific; UMIC = upper-middle-income country.

MAP 3.1   Use of fintech credit around the world, 2014–17

INDEX SCORE:
- 0–5
- 5–18
- 18–35
- 35–80
- 80–100
- No data

IBRD 48278 | AUGUST 2024

Source: Didier et al. 2022.
Note: This map is based on an index on the usage of fintech credit, which was constructed using a single indicator: the value of total new financing intermediated through electronic platforms, accumulated over the period 2014–17 (measured as a percentage of GDP). The index covers 173 countries from all geographical regions and across all income groups. Fintech = financial technology.

to regulators and supervisors (Crisanto, Ehrentraud, and Fabian 2021; FSB 2019; IMF 2019).

Traditional financial service providers are also embracing and emulating digital service delivery. A recent survey of fintech market participants[5] reveals that the future of the financial sector will likely combine physical and digital ("phygital") aspects. Across regions one message was clear: digitalization does not spell the end of physical infrastructures for financial services. However, compared to regions like the Middle East and North Africa (66 percent) and Europe and Central Asia (61 percent), the EAP region (38 percent) has fewer traditional financial institutions that believe physical branches will dominate financial service delivery, which signifies a greater intent to adapt to the ongoing changes. Out of the box 3.3 focuses on the experience of Cambodia and the introduction of digital payments by a commercial bank owned by a foreign bank.

The rapid expansion of digital financial services gives rise to important policy trade-offs. The fast pace and disruptive nature of these technological advances and emerging business models can make it more challenging for policy makers and regulators to ensure that market outcomes remain aligned with core policy objectives. At lower levels of fintech development, providing basic policy support for innovation and mitigating immediate risks, such as illicit activity and protection of customer funds, may yield good short-term outcomes as policy makers aim to reap innovation, inclusion, and efficiency gains. Policy makers, however, should be mindful that adoption can increase rapidly, which will require them to improve their monitoring tools and be ready to step in. Strengthening or clarifying policy frameworks and improving financial infrastructures become increasingly important to

Out of the box 3.3.   Digital financial services in Cambodia

Digital technologies are transforming Cambodia's financial landscape by expanding access to services and fostering financial inclusion. After significant reforms in the 1990s, the banking system is now both highly liberalized and dollarized. Cambodia has seen an influx of banks, mostly foreign, looking to establish themselves in a crowded market. Microfinance institutions are also prevalent and hold an important place in the financial ecosystem, both as large deposit takers and as lenders. Together, banks and microfinance institutions are the driving agents of Cambodia's digital financial ecosystem.

Despite a widespread surge in account ownership (from 4 percent in 2011 to 33 percent in 2021), only 7 percent of adults save at a financial institution (figure OB3.3.1, panel a). This disconnect between financial access and formal savings inhibits the efficient allocation of credit (World Bank 2021). Geographical hurdles are central to the high costs of embracing formal savings. About one in three Cambodian adults grapples with long distances to financial institutions, with more than half the population living over 30 minutes away from the nearest ATM, bank, or microfinance institution (Beed Management 2017).

Well-designed and implemented digital financial services can help bridge both physical and social divides to expand access for historically underbanked populations, bolster formal savings, and improve access to credit. Fortunately, the banking sector has caught the digital bug and, with government support, is capitalizing on Cambodia's high mobile phone penetration to bring financial services closer to customers (figure OB3.3.1, panel b).

FIGURE OB3.3.1   Cambodia has witnessed a gradual increase in access to formal financial services, with digital payments increasing in popularity but formal savings remaining low
Adults using formal financial services and digital payments

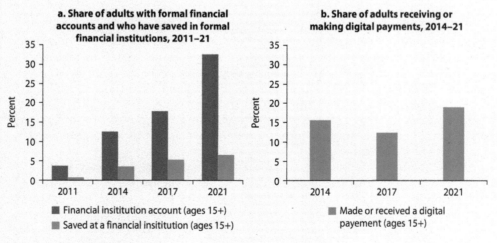

Source: Global Findex Database, 2021.

(*Continued*)

Out of the box 3.3.   (Continued)

ABA Bank, a subsidiary of the National Bank of Canada, became Cambodia's largest commercial bank in 2022, and has contributed to increasing digitalization. The following comments are excerpts from a conversation with Dominic Jacques (Deputy Vice President, International Development, National Bank of Canada, and Member of the Board of Directors at ABA Bank).

What was your experience with the entry into the Cambodian market?

The objective of our international strategy was for our shareholders to benefit from the strong economic growth in developing/emerging markets. The way we decided to do that was to invest in banks in these markets, be an active shareholder, help them grow and contribute to the local economy. We were open to either minority or majority participations. We were looking for countries with high economic growth, favorable demographics, and underbanked populations. Cambodia met all the criteria. The GDP growth was very high, historically, 6 to 7 percent per year. The population with access to a bank account was probably below 15 percent, which represented a very interesting growth potential. So, we were pleased to invest in Cambodia. Cambodia has no limitations on foreign investments into the banking sector. This allowed us to put an initial investment into ABA Bank, with a pathway to increase our stake over time. We acquired a controlling stake in 2016 and increased it to 100 percent ownership in 2018.

What role has technology played in ABA's operations in Cambodia?

At the start of our operations, the economy was heavily cash based. For example, you would have someone who was going to buy a car go to the bank across from the car dealership and withdraw cash equal to the full price of the car. A little later, the car dealer would carry the cash across the street to deposit the money in an account in the same bank. The low adoption of digital financial services was due to low levels of digital infrastructure and high costs of traditional digital finance products like debit cards and points of sales terminals. There was also a lack of trust in digital finance products, which were harder for people to understand. Recognizing the gap, our experienced management team at ABA developed a QR code–based payment in Cambodia. The technology was already popular in other Asian markets and Cambodia quickly embraced the innovation, partly because of its young and tech-savvy population. We also saw the level of digital transactions skyrocket during COVID, accelerating the shift that had already begun.

safely supporting fintech adoption, as fintech scales up, reaches more consumers, and increases its dependence on larger volumes of user data—an issue elaborated on in more detail in the next chapter.

Notes

1. Establishing a causal link is admittedly harder because of the difficulty in disentangling the effect of digitalization on productivity from the effect of productivity on digitalization adoption.

2. Refer also to Cusolito, Lederman, and Pena (2020) for evidence from developing countries on the positive productivity effects of digitalization in the manufacturing sector.
3. These qualifications are reminiscent of the findings of early studies on the impact of information and communication technology adoptions, such as Bugamelli and Pagano (2004).
4. As explained in the preface, this report includes a new feature named "out of the box," which spotlights specific country-sector cases designed to help the reader to better appreciate the themes addressed in the report.
5. The Fintech Market Participants Survey under the Future of Finance series 2022 (World Bank 2022).

References

Arnold, J. M., B. Javorcik, M. Lipscomb, and A. Mattoo. 2016. "Services Reform and Manufacturing Performance: Evidence from India." *Economic Journal* 126 (590): 1–39.

Autor, D., 2024. "Applying AI to Rebuild Middle Class Jobs." NBER Working Paper 32140, National Bureau of Economic Research, Cambridge, MA.

Bailin Rivares, A., P. Gal, V. Millot, and S. Sorbe. 2019. "Like It or Not? The Impact of Online Platforms on the Productivity of Incumbent Service Providers." OECD Economics Department Working Paper 1548, OECD Publishing, Paris.

Barnett, K., S. W. Mercer, M. Norbury, G. Watt, S. Wyke, and B. Guthrie. 2012. "Epidemiology of Multimorbidity and Implications for Health Care, Research, and Medical Education: A Cross-Sectional Study." *Lancet* 380 (9836): 37–43.

Beed Management. 2017. "Making Access Possible: Cambodia Diagnostic Report 2017." Centre for Financial Regulation and Inclusion, FinMark Trust, and United Nations Capital Development Fund. https://finmark.org.za/system/documents/files/000/000/222/original/Cambodia_Diagnostic_30-October-2017.pdf?1601985461.

Bloom, D. E., and D. Canning. 2008. "Population Health and Economic Growth." Commission on Growth and Development Working Paper 24, World Bank, Washington, DC.

Brancati, U., M. Cesira, A. Pesole, and E. Fernández Macías. 2020. *New Evidence on Platform Workers in Europe*. EUR 29958 EEN. Luxembourg: Publications Office of the European Union.

Brynjolfsson, E., X. Hui, and M. Liu. 2019. "Does Machine Translation Affect International Trade? Evidence from a Large Digital Platform." *Management Science* 65 (12): 5449–60.

Brynjolfsson, E., D. Li, and L. Raymond. 2023. "Generative AI at Work." NBER Working Paper 31161, National Bureau of Economic Research, Cambridge, MA.

Bugamelli, M., and P. Pagano. 2004. "Barriers to Investment in ICT." *Applied Economics* 36 (20): 2275–86.

Cazzaniga, M., F. Jaumotte, L. Li, G. Melina, A. Panton, C. Pizzinelli, E. Rockall, and M. Tavares. 2024. "Gen-AI: Artificial Intelligence and the Future of Work." IMF Working Paper 2024/001, International Monetary Fund, Washington, DC.

Chen, G., C. K. K. Chan, K. K. H. Chan, S. N. Clarke, and L. B. Resnick. 2020. "Efficacy of Video-Based Teacher Professional Development for Increasing Classroom Discourse and Student Learning." *Journal of the Learning Sciences* 29 (4–5): 642–80.

CIPD (Chartered Institute of Personnel and Development). 2017. "To Gig or Not to Gig? Stories from the Modern Economy." CIPD, London.

Crisanto, J. C., J. Ehrentraud, and M. Fabian. 2021. "Big Techs in Finance: Regulatory Approaches and Policy Options." FSI Brief No. 12, Financial Stability Institute, Bank for International Settlements, Basel.

Cusolito, A., D. Lederman, and J. Pena. 2020. "The Effects of Digital-Technology Adoption on Productivity and Factor Demand: Firm-Level Evidence from Developing Countries." Policy Research Working Paper 9333, World Bank, Washington, DC.

David, P. 1990. "The Dynamo and the Computer: An Historical Perspective on the Modern Productivity Paradox." *American Economic Review* 80 (2): 355–61.

Demirgüç-Kunt, A., L. Klapper, D. Singer, and S. Ansar. 2022. *The Global Findex Database 2021: Financial Inclusion, Digital Payments, and Resilience in the Age of COVID-19*. Washington, DC: World Bank.

Didier, T., E. Feyen, R. Llovet Montanes, and O. Ardic. 2022. "Global Patterns of Fintech Activity and Enabling Factors." Fintech and the Future of Finance Flagship Technical Note, World Bank, Washington, DC.

Duggan, V., S. Rahardja, and G. Varela. 2013. "Service Sector Reform and Manufacturing Productivity: Evidence from Indonesia." Policy Research Working Paper 6349, World Bank, Washington, DC.

Farrell, D., F. Greig, and A. Hamoudi. 2018. "The Online Platform Economy in 2018: Drivers, Workers, Sellers and Lessors." JPMorgan Chase Institute, New York.

Felten, E. W., M. Raj, and R. Seamans. 2023. "Occupational Heterogeneity in Exposure to Generative AI." https://papers.ssrn.com/sol3/papers.cfm?abstract_id=4414065.

Frey, C. B., and M. A. Osborne. 2017. "The Future of Employment: How Susceptible Are Jobs to Computerisation?" *Technological Forecasting and Social Change* 114: 254–80. Elsevier.

FSB (Financial Stability Board). 2019. "BigTech in Finance: Market Developments and Potential Financial Stability Implications." FSB, Basel.

FSO (Switzerland, Federal Statistical Office). 2020. "Internet-Mediated Platform Work Is Not Very Common in Switzerland." Press Release, May 19, 2020. https://www.bfs.admin.ch/bfs/en/home.assetdetail.12787865.html.

Gal, P., G. Nicoletti, T. Renault, S. Sorbe, and C. Timiliotis. 2019. "Digitalisation and Productivity: In Search of the Holy Grail—Firm-Level Empirical Evidence from European Countries Economics Department." OECD Economics Department Working Paper 1533, OECD Publishing, Paris.

Google, Temasek, and Bain & Company. 2022. "E-Conomy SEA Report 2022." Google, Temasek, and Bain & Company. https://services.google.com/fh/files/misc/e_conomy_sea_2022_report.pdf?utm_source=bain&utm_medium=website&utm_campaign=2022.

IHME (Institute for Health Metrics and Evaluation). 2023. "Global Burden of Disease 2019." Hans Rosling Center for Population Health, Seattle, WA.

ILO (International Labour Organization). 2021. *World Employment and Social Outlook 2021: The Role of Digital Labour Platforms in Transforming the World of Work*. Geneva: International Labour Office.

Ilsøe, A., and L. Weber Madsen. 2017. "Digitalisation of the Labour Market: Danish Experience with Automation and Digital Platforms." Employment Relations Research Center (FAOS), University of Copenhagen.

IMF (International Monetary Fund). 2019. *Fintech: The Experience So Far*. IMF Policy Paper No. 2019/024. Washington, DC: IMF.

Ito, H., K. Kasai, H. Nishiuchi, and M. Nakamuro. 2019. "Does Computer-Aided Instruction Improve Children's Cognitive and Non-Cognitive Skills? Evidence from Cambodia." RIETI Discussion Paper 19040, Research Institute of Economy, Trade and Industry.

Kanazawa, K., D. Kawaguchi, H. Shigeoka, and Y. Watanabe. 2022. "AI, Skill, and Productivity: The Case of Taxi Drivers." NBER Working Paper 30612, National Bureau of Economic Research, Cambridge, MA.

Li, H., Z. Liu, F. Yang, and L. Yu. 2023. "The Impact of Computer-Assisted Instruction on Student Performance: Evidence from the Dual-Teacher Program." IZA Discussion Paper No. 15944, Institute of Labor Economics, Bonn.

Mo, D., Y. Bai, Y. Shi, C. Abbey, L. Zhang, S. Rozelle, and P. Loyalka. 2020. "Institutions, Implementation, and Program Effectiveness: Evidence from a Randomized Evaluation of Computer-Assisted Learning in Rural China." *Journal of Development Economics* 146: 102487.

Noy, S., and W. Zhang. 2023. "Experimental Evidence on the Productivity Effects of Generative Artificial Intelligence." *Science* 381 (6654): 187–92.

OECD (Organisation for Economic Co-operation and Development). 2020. "Realising the Potential of Primary Health Care." OECD Health Policy Studies, OECD, Paris.

Pizzinelli, C., A. Panton, M. M. Tavares, M. Cazzaniga, and L. Li. 2023. "Labor Market Exposure to AI: Cross Country Differences and Distributional Implications." IMF Working Paper 2023/216, International Monetary Fund, Washington, DC.

SF (Statistics Finland). 2017. *Labour Force Survey: Platform Jobs 2017*. Helsinki: SF.

Shi, L., M. Makinen, D. Lee, R. Kidane, N. Blanchet, H. Liang, J. Li, M. Lindelow, H. Wang, S. Xie, and J. Wu. 2015. "Integrated Care Delivery and Health Care Seeking by Chronically-Ill Patients—A Case-Control Study of Rural Henan Province, China." *International Journal for Equity in Health* 30 (14): 98.

Trammell, P., and A. Korinek. 2023. "Economic Growth under Transformative AI." NBER Working Paper 31815), National Bureau of Economic Research, Cambridge, MA.

WHO (World Health Organization). 2018. "Building the Economic Case for Primary Health Care: A Scoping Review." WHO/HIS/SDS/2018.48, May 8, 2018, WHO, Geneva.

World Bank. 2021. "Resilient Development: A Strategy to Diversify Cambodia's Growth Model." Cambodia Country Economic Memorandum, World Bank, Washington, DC.

World Bank. 2022. "World Bank Group Global Market Survey: Digital Technology and the Future of Finance." Fintech and the Future of Finance Technical Note, World Bank, Washington, DC.

World Bank. 2023. "Survey on Preparedness and the Situation for Health Information Technology in the EAP Countries." World Bank, Washington, DC.

World Bank. Forthcoming. *Bridging the Basic Learning Gap: Supporting Teachers in Middle-Income East Asia and the Pacific*. Washington, DC: World Bank.

Worldpay. 2022. "The Global Payments Report 2022." Worldpay, Cincinnati, OH.

Yarrow, N., C. Abbey, S. Shen, and K. Alyono. Forthcoming. *Using Education Technology to Improve K–12 Student Learning in East Asia and Pacific: Promises and Limitations*. Washington, DC: World Bank.

Zhang, Z., S. Mishra, E. Brynjolfsson, J. Etchemendy, D. Ganguli, B. Grosz, T. Lyons, J. Manyika, J.C. Niebles, M. Sellitto, Y. Shoham, J. Clark, and R. Perrault. 2021. "The AI Index 2021 Annual Report." AI Index Steering Committee, Human-Centered AI Institute, Stanford University, Stanford, CA.

What Should Be Done? 4

Abstract

To ensure that services development is inclusive and sustainable, countries in the East Asia and Pacific (EAP) region must strike a balance between (1) liberalization and regulation, (2) the state and the market in providing infrastructure and education, and (3) unilateral domestic reform and cooperative international action.

Balanced policy action

So far, this book has argued that the powerful combination of technological change and policy reforms can generate enhanced opportunities, through higher productivity and changing demand for skills, as well as through higher capacities from greater access to health, education, and finance. To unleash a virtuous cycle between opportunity and capacity (depicted at the bottom of this book's organizing framework in figure 1.9), and to ensure that services development is inclusive and sustainable, countries in the EAP region must take three pairs of policy actions.

- First, they must strike a balance between liberalization and regulation. That balance requires removing the many policy distortions that inhibit entry and competition in their services markets while instituting a regulatory framework that addresses old and new market distortions, including the distortive concentration of market power and data abuse that can arise in markets where digital platforms dominate.
- Second, they must strike a balance between the state and the market in creating the infrastructure and skills needed to take advantage of the opportunities that are emerging. Over the last decades, the democratization of access to mobile telephony provided by competing private firms seemed to have obviated the need for the fixed-line networks created by plodding public sector monopolies. The digital benefits of access to high-speed broadband, however, have revived the question of how the state can ensure adequate access for the poor and remote.

Countries must grapple with the challenge of determining the extent to which the market and private institutions can be relied on to deliver the skills needed by the digital services economy, and what role the state must play in ensuring the appropriate scale, composition, quality, and access.

- Third, EAP and other countries must strike a balance between unilateral domestic reform and cooperative international action to address services market failures that have a transborder dimension. One example is the need to ensure that heterogeneity in national regulatory approaches to privacy and cybersecurity do not impede the data flows central to the global services economy. Another example is the need to ensure that international transportation, central to global trade and tourism, does not continue to add carbon dioxide emissions to the atmosphere. In both these cases, countries are beginning to cooperate meaningfully in both regional and multilateral forums.

The chapter turns now to the discussion of each of these three pairs of policy actions.

Balancing liberalization and regulation

EAP countries must create competitive conditions in which consumers of services benefit. Doing so requires dealing with both policy and market distortions.

The unfinished business of services liberalization

Despite the progress of reforms in EAP services sectors described in previous chapters, new data from the World Bank–World Trade Organization Services Trade Restrictiveness Index for 2022 reveal that it is still an unfinished business. As of 2022, EAP countries still have relatively restrictive regimes for services trade in most sectors (figure 4.1, panel a). Moreover, compared to other countries, EAP countries have higher restrictiveness than would be expected by their level of development (figure 4.1, panel b).

To further examine the specific policy measures behind the numerical values for each sector and for each mode of delivery, the analysis first considers the restrictions applied to cross-border trade (mode 1) and commercial presence (mode 3), because they are the most prevalent for the subsectors considered. The tables in appendix G report the main restrictions applied by the larger EAP countries in different sectors. The most prevalent restriction to cross-border trade (mode 1) is the commercial presence requirement, namely the need to establish a branch or an affiliate to be able to serve the market (table G.1). The most common restrictions applied to the establishment of a commercial presence are maximum percentage allowed of foreign equity and discriminatory licensing (table G.2).

For instance, Malaysia, Thailand, and Viet Nam restrict foreign ownership in telecommunications and maritime transportation, and Malaysia and Thailand also do so in commercial banking and insurance. Indonesia requires joint ventures in banking and insurance. Many countries also apply "economic needs tests," that is, they issue new licenses only when they feel they need new firms and do so in a way that is not transparent or based on objective criteria. Several countries, such as Indonesia, Malaysia, Thailand, and Viet Nam, restrict cross-border flows of

FIGURE 4.1 Most EAP countries restrict services trade more than other economies at comparable levels of development

STRI values in selected EAP countries, by sector and GDP per capita, 2022

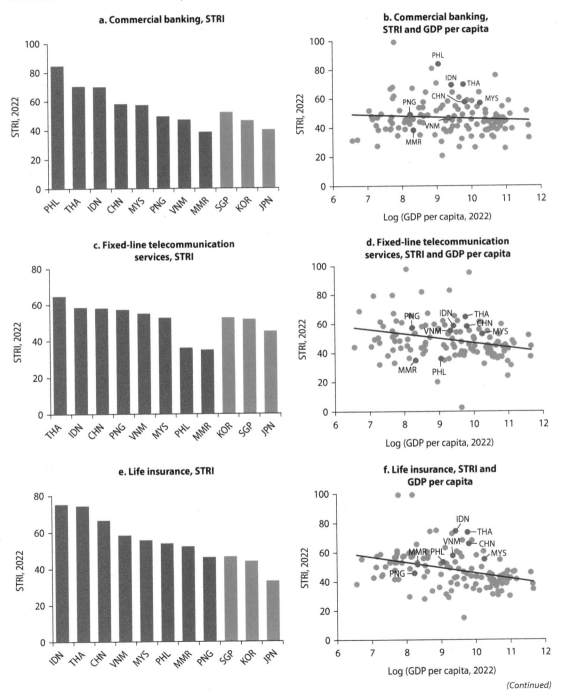

(Continued)

FIGURE 4.1 Most EAP countries restrict services trade more than other economies at comparable levels of development *(Continued)*

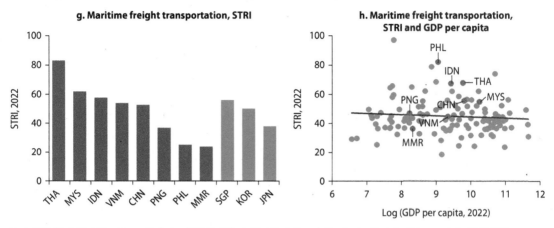

g. Maritime freight transportation, STRI

h. Maritime freight transportation, STRI and GDP per capita

Source: World Bank, World Development Indicators, 2022; World Bank–World Trade Organization, Services Trade Restrictiveness Index (STRI), 2022.
Note: The figure reports STRI values for the indicated services sectors in the larger EAP countries, as well as for four advanced economies: Hong Kong SAR, China; Japan; the Republic of Korea; and Singapore. It also presents accompanying scatter plots for each sector, featuring the STRI Index for the sectors against the level of development (measured as the logarithm of the GDP per capita). EAP = East Asia and Pacific.

financial, transportation, and other services, and impose strict conditions on international data flows.

Two recent policy developments in Indonesia and the Philippines highlight some progress in closing the services reform gaps: (1) the Financial Sector Omnibus Law (FSOL) in Indonesia, and (2) the amendment to the Public Service Act (PSA) in the Philippines. The following paragraphs provide more information on these reforms.

After almost two years of preparation, Indonesia's FSOL passed in early 2023. The FSOL integrates 17 institutional and sectoral laws underpinning the financial sector and includes key institutional architecture reforms in the areas of financial stability, long-term finance, sustainable finance, financial innovation, consumer protection, and access to finance. It complements both the Omnibus Law on Job Creation of 2021 and the Health Omnibus Law of June 2023.

The FSOL is expected to promote the deepening of the financial sector, with increased availability and diversification of financial products and services. It will facilitate this deepening through (1) the mandatory provision of commercial bank loans to underserved sectors and areas (for example, micro, small, and medium enterprises; inclusive financing; and sustainability financing); (2) new emerging financial products (for example, carbon trading, bullion for gold investment, and crypto); (3) expansion of general insurance and life insurance business scope; and (4) additional access through various providers, including cooperatives and microfinance institutions.

In the Philippines, until recently, ambiguity in the legal definition of what constitutes a "public utility" has restricted foreign entry into certain services sectors—with implications for competitiveness. The recent amendment of the PSA addresses this ambiguity, opening key enabling services sectors to increased competition. Whereas some restrictions remain for specific public services, the 40 percent foreign ownership cap (provided for in the Philippine Constitution) for public utilities no longer applies to public services other than critical infrastructure. Notably, the amended PSA would allow 100 percent foreign ownership in the following public services now excluded from the definition of public utility: telecommunications, domestic shipping, airlines, railways and subways, expressways, and tollways.[1]

The PSA amendment is a continuation of a foreign direct investment liberalization process taking place in the Philippines. Other recent liberalizations include the Retail Trade Liberalization Act (December 2021), the Foreign Investment Act (March 2022), and the Renewable Energy Act (November 2022).

A fully and effectively implemented PSA amendment is estimated to increase total factor productivity by 3.2 percent on average, and up to 6.4 percent in electrical machinery–related sectors that rely intensively on telecommunication and transportation inputs. Appendix H presents an analysis of the potential economic effects of the amendments to the PSA.

The PSA amendment also represents an interesting case in terms of political economy of reforms. The economic cabinet secretaries championed the reform with the aim of jumpstarting investments for an economy that was hit hard by pandemic lockdowns and by a decelerating global economy. The timing was opportune, and these efforts were met by a receptive legislature eager to demonstrate bold action to boost growth. In its last remaining months in power, the previous administration successfully marshalled its majority coalition in the Philippine Congress to pass this landmark liberalization reform. To address concerns that the reform would endanger national security by allowing foreigners to assume control of key sectors, the measure empowered the president to prohibit investments in public services in the interest of national security.

The National Economic Development Authority issued the Implementing Rules and Regulations of the amended PSA in April 2023. As full implementation materializes, beyond the productivity effects, the reform might also contribute to lower food prices because food-producing sectors rely on transportation services for their supply chain management. Lower prices, in turn, might increase public support for the reforms.

In conclusion, the evidence reported in figure 4.1 calls for further liberalization of services trade in EAP countries. However, even when policy reforms that liberalize services sectors are implemented, uneven implementation across subnational regulatory authorities or regions could undermine the effects of those reforms. "Out of the box" 4.1 explores these issues using the case of the logistics sector in Indonesia.

Finally, annex 4A at the end of this chapter examines more specifically the state of restrictions in services trade for the Pacific islands, highlighting how liberalization has particular importance for countries characterized by small market sizes and remoteness.

The emerging business of regulating the digital economy

The digital services economy gives rise to new sources of market failures and regulatory challenges. Policy makers will need to strengthen regulation to ensure that digital

Out of the box 4.1.   Logistics services in Indonesia

Logistics services play a pivotal role in driving firm productivity and improving market access in Indonesia, the world's largest archipelagic nation. For Indonesian businesses looking to scale up domestically and internationally, efficient and reliable logistics services facilitate supply chain management and encourage businesses to join global value chains. Over the years, the government has reformed policy to facilitate the entry and operations of foreign and domestic logistics firms, and costs have declined (figure OB4.1.1). Although challenges remain, logistics firms are increasingly deploying technology to provide better services.

FIGURE OB4.1.1   Logistics costs have steadily declined, reflecting, in part, changes in policy and improvements in logistics services
National logistics costs per GDP, Indonesia, 2005–21

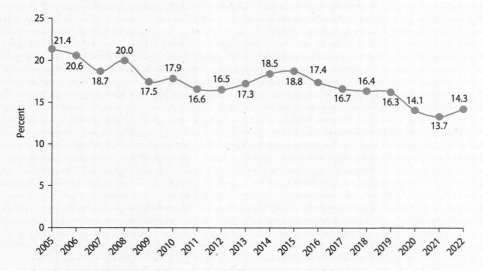

Source: Indonesia, Ministry of National Development Planning (BAPPENAS) 2023.

In interviews, experts in logistics policy, in logistics service provision, and from firms benefiting from logistics services offered the following insights.

What is the current regulatory landscape, and how has it affected the logistics services sector in Indonesia?

Lamiaa Bennis, Logistics Sector Specialist, at the World Bank: There has been notable progress in logistics sector policy over the years. In 2012, the government passed the National Logistics Blueprint that outlined investment and licensing reforms and prioritized public investment projects to expand transport networks. In alignment with the blueprint, the Indonesian government implemented a series of reforms to enhance (1) port governance and efficiency, (2) the competitiveness of the logistics services sector, and (3) the efficiency of cross-border processes. During 2016 and 2017, the government removed conditions that hindered private sector participation in commercial ports. Furthermore, in 2021, the four major

(Continued)

Out of the box 4.1. (Continued)

state-owned port operators, collectively known as Pelindo I to IV, underwent a comprehensive restructuring and were consolidated under a single entity called Pelindo. This strategic merger is anticipated to significantly enhance overall port productivity and efficiency, thereby benefiting stakeholders in the logistics industry. The Omnibus Law on Job Creation in 2021 helped remove foreign equity restrictions in freight forwarding, storage, and warehousing, and in cargo handling. However, foreign ownership in courier services and most modes of freight transportation (except rail) remains limited to 49 percent.

Despite these reforms, regulatory oversight of the logistics industry continues to be highly fragmented. Freight forwarding services, maritime transport services, air transportation services, rail transportation services, and freight transportation are under the Ministry of Transportation, while warehousing and storage are under the Ministry of Trade. Customs is under the Ministry of Finance, which is typical, and e-commerce logistics companies, courier, and express logistics companies are under the Ministry of Information and Telecommunications. So, a third-party logistics provider company seeking to offer freight forwarding and warehousing services would need to comply with regulations under both the Ministry of Transportation (for freight forwarding) and the Ministry of Trade (for warehousing). In some cases, there are also differences between national-level regulations and sub-national-level regulations, such as for trucking services. The logistics industry needs an environment that enables coordination and prevents disruption to ensure the realization of its full potential.

Given the unique geographical challenges Indonesia faces, what are some of the day-to-day issues logistics firms face in delivering goods within and outside of Indonesia?

Trismawan Sanjaya, Vice Chair, Indonesian Logistics and Forwarders Association: Indonesia still has gaps in infrastructure, especially for inland transportation and in seaport infrastructure, which is especially challenging given that Indonesia is an archipelago. In addition, many manufacturing industries in the west of the country deliver to the east but their logistics partners face challenges in selling container space for the return journey. Logistics regulations and policies, especially across the different islands, are also not standardized, adding further to costs and uncertainties.

I would also highlight the challenge of financing. The need for collateral is a major issue for asset-light companies such as freight forwarders and third-party logistics providers. Although overall interest rates have been on the decline, Indonesia still has high lending rates relative to its regional peers—in some cases more than double. Logistics customers also demand longer payment terms, which doubled from 30 to 60 days after the pandemic. Such long payment terms put additional pressure on logistics companies that need working capital to pay their employees and subcontractors. Access to finance therefore has an impact on the competitiveness of logistics services providers in Indonesia in general and puts domestic companies at a disadvantage compared to foreign companies that have access to finance on more favorable terms.

(Continued)

Out of the box 4.1.   (Continued)

How have digitalization and technological change improved the quality of/ access to logistics services in Indonesia? What specific technologies have been transformational?

Tiger Fang, Co-founder and CEO at Kargo Technologies: In Indonesia, logistics and transportation providers are very fragmented. Over 95 percent of truck owners own less than five trucks and operate independently. In that sense, technologies to improve the access to those providers by using managed marketplace technology have been game changing. By providing a platform of vetted vendors and customers, logistics companies like Kargo have been able to transform the way firms look for logistics support, and have supported improved earnings, better working capital, and profitable growth. In our case, our enterprise Multinational Company customers use our order management, track and trace, document reconciliation, and contract/fleet management modules to enable digitized workflow for their operators. These technologies allow for more competitive bidding across nationwide vendors, better cash allocations for treasury, and also streamlined document reconciliation.

firms respect prudential concerns (for example, in financial markets), address data protection and cybersecurity risks. and do not engage in anticompetitive behavior to the detriment of their consumers and workers.

Digital platforms

Although they can generate significant economic benefits in terms of productivity, jobs, and access, digital platform firms also present risks associated with market concentration and data security and privacy. Lack of product market competition and misuse of data (commercially motivated surveillance) to generate rents are among the most salient concerns related to the platform revolution. Without regulation, the expected benefits of platform firms can reverse course. In the short run, digital platforms can generate rents at the expense of decreasing consumers' surplus or gains for small suppliers, as well as business opportunities for innovators. In the long run, digital platforms can discourage innovation (figure 4.2).[2]

Risks: Weak competition

Risks to competition by digital platforms vary by platform business model, stage in the platform life cycle, and digital maturity of the country. On all three fronts, digital platforms in EAP have considerable country variation, particularly with respect to regulations (for example, on competition and data) and country income level. Moreover, weaker competitive pressures affect a variety of actors—consumers, suppliers, and competitors—in different ways.

Risks to competition are primarily associated with entry barriers and rivalry constraints that arise from network effects, self-regulation, and data collection. Direct network effects (benefits to consumers on the same side of the network) and indirect

FIGURE 4.2 **Digital platforms come with benefits and risks**

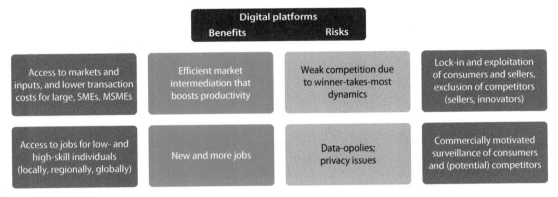

Source: World Bank 2023.
Note: Other economic and social issues include taxation, labor, and social protection, and the psychological effects of social network platforms, political polarization, fake news, and freedom of speech. MSMEs = micro, small, and medium enterprises; SMEs = small and medium enterprises.

network effects (when the value of the platform to one group depends on how many members of another group participate) lead to monopolistic (that is, winner-take-all or winner-take-most) structures. Dominant players emerge because of a combination of these network effects, their ability to set the rules of the game for interactions on the platform, and access to data. In addition, scale may translate to greater volume of data, and scope may increase variety of data. Both help accelerate existing feedback loops that further strengthen the dominant player.

Consumers face risk of lock-in and data exploitation when dominant platforms emerge. Customer lock-in may be driven by direct network effects (enhanced platform experience as the number of users grows) and indirect network effects (larger platforms attracting more third-party participants). Network effects reap the greatest benefits at scale economies (number of users and volume of data), which are then difficult to transfer or multihome across networks. Dominant networks can enforce switching costs or even barriers by limiting data portability and interoperability, which raises additional concerns over data privacy and access for users.

Suppliers also face lock-in risks as well as self-preferencing risks posed by vertically integrated platform operators. Dominant platforms may harm third-party sellers and advertisers as economies of scope begin to take shape in forming a dominant ecosystem. The platform operator serves as a nexus of multiple party interactions, through which it can exploit market intelligence asymmetries. In addition to information asymmetries, platforms may simply engage in self-preferencing over competing participants or restricting alternative options for their users. For example, in 2018, the Fair Trade Commission in Japan ruled that Airbnb restricted listings of houses in platforms other than Airbnb.

These challenges associated with digital markets have triggered antitrust enforcement (figure 4.3). As of September 2022, competition authorities in EAP had finalized 20 abuse of dominance cases and 12 anticompetitive agreements, with only 9 cases in middle-income countries (China, Indonesia, Malaysia, the Philippines, and Viet Nam). The World Bank's digital antitrust database confirms that competition

FIGURE 4.3   Challenges associated with digital markets have triggered antitrust enforcement

Digital antitrust cases, by region, 2023

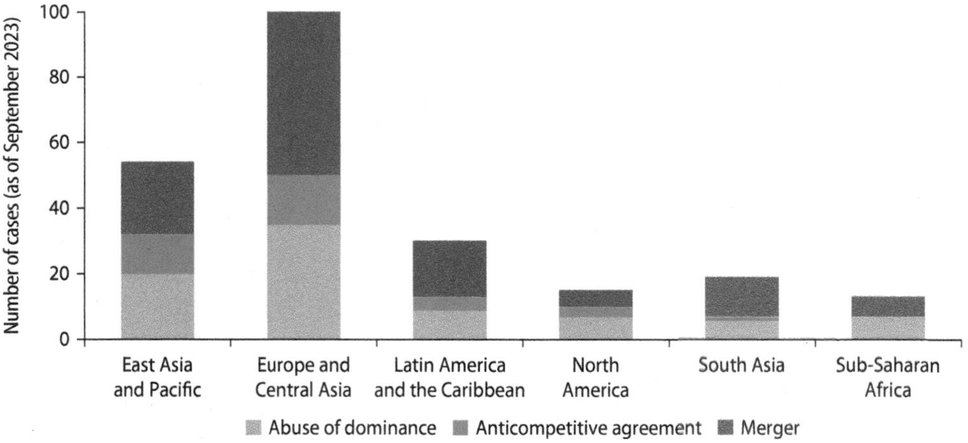

Source: World Bank, Global Digital Antitrust Database, updated as of September 2023.

cases in the region have mostly focused on interplatform competition and pricing in high-income countries as compared to offline versus online competition in middle-income countries. For instance, the Malaysia Competition Commission sanctioned the company providing access to the National Single Window for requiring users to access it through the company's own electronic mailbox. Regarding licensing practices, the Korean Fair Trade Commission concluded in 2021 that Google had banned manufacturers from releasing devices that run on rival operating systems as a prerequisite for licensing Play Store. Finally, that commission sanctioned Naver, a popular search platform, for altering its algorithm to preference products offered on its own "open market." This behavior harmed nonassociated vendors whose products were demoted to lower positions on Naver's ranking. The integration of Naver into different markets was key for its ability to self-preference.

Digital platforms in the region have also used mergers and acquisitions, often targeting innovating rivals, to reinforce positions of dominance and expand into digital conglomerates. Content generation and software design are the top digital industries targeted by conglomerate acquirers. Global digital platforms, including Amazon, Apple, Facebook, Google, and Microsoft, have undertaken an intense merger strategy, often through conglomerate mergers, the earliest of which flew under the radar of antitrust authorities or benefited from lighter reviews (Motta and Peitz 2021). Among all regions, EAP has seen the highest number of conglomerate mergers and acquisitions, with a growing share of digital mergers each year between 2012 and 2021. Over the same period, digital mergers tripled in the Republic of Korea and rose by 60 percent in the remaining EAP economies (figure 4.4).[3] As of September 2022, however, competition authorities in EAP had reviewed only 22 mergers. Lack of ex ante merger control frameworks and the ability of digital mergers often to remain below merger notification thresholds

FIGURE 4.4   **Evolution of conglomerate digital mergers, 2012–21**

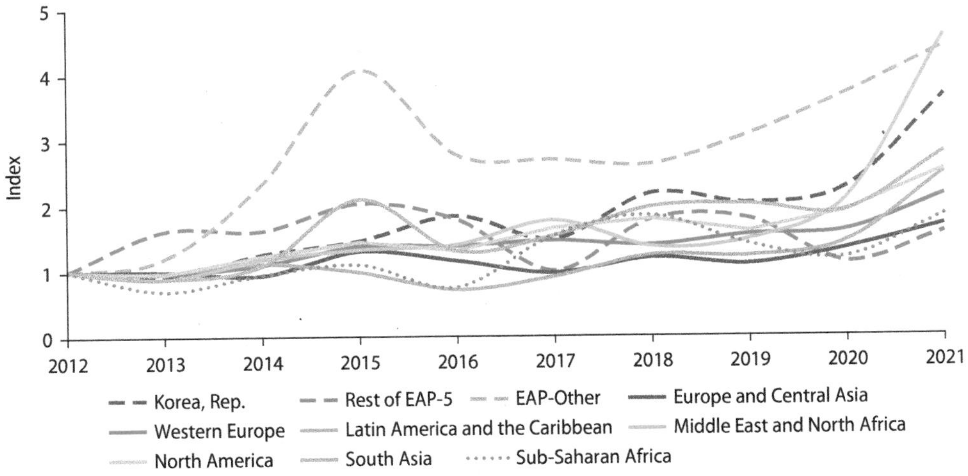

Source: World Bank, forthcoming.
Note: EAP = East Asia and Pacific.

have prevented countries from reviewing mergers with likely anticompetitive effects and from setting appropriate remedies. The acquisition of Uber by Grab exposed these challenges, including notification thresholds that did not capture relevant concentration operations, delayed resolutions, and limited coordination between competition authorities in the region. Even though it brought together the largest two players in ride-hailing, the transaction did not require ex ante merger notification in any of the economies involved.

Risks: Data-opolies and privacy issues
Personal data lie at the heart of digital platforms and contain inherent market value for platform optimization and third parties such as advertisers. User data come with unique responsibilities pertaining to privacy and morality. Threats of misuse for commercial purposes, exploitation, and cybersecurity are some of the most prominent risks to users. The value of market intelligence and interest by third parties can incentivize misuse. Conflicting challenges also emerge with respect to benefits of data mobility and interoperability for fostering competition, versus ensuring consumer privacy and security.

Market intelligence on competitor insights and consumer data intersect with competition concerns. Digital platforms may use data to exploit both customers and competitors (in the case of vertically integrated platforms). Digital platforms may engage in price fixing by algorithms to exploit consumer willingness to pay. At the same time, platforms may use their superior market intelligence data arising from buyer-seller interactions to unfairly outperform competitors.

Evidence exists of positive feedback loops between pro-competitive efforts and data privacy. Digital platforms have an incentive to collect as much data as possible, both for optimizing their own services for users' benefit and for selling to third parties. In contestable markets, with nonmyopic consumers, market players have an incentive to retain consumer data in the most responsible manner possible—otherwise, they risk losing participants. Concerns about autonomy, service quality, and privacy are often addressed to retain consumers.[4]

EAP authorities have already started investigating cases related to personal data protection and data management involving digital platforms. As of September 2022, 32 cases of infringements of data protection rules involving digital platforms had been analyzed in China, Korea, the Philippines, and Singapore (Barzelay 2022), and competition authorities had analyzed data transparency and data as a barrier to entry in 6 abuse of dominance cases.

Several EAP economies have approved proposals for regulation of digital platforms in recent years. Japan in 2020 approved the Act on Improving Transparency and Fairness of Specific Digital Platforms. Australia approved the News Media and Digital Platforms Mandatory Bargaining Code in 2021 and has proposed developing codes for specific types of platforms such as retail. Thailand recently adopted a law requiring registration of digital platforms, and Korea has proposed the Act on Fairness in Online Platform Intermediary Transactions. Viet Nam is drafting a law to regulate platforms.

Looking ahead, the main challenges to ensuring competitive dynamics relate to adapting competition frameworks to the realities of the digital economy. Antitrust frameworks for abuse of dominance, agreements among competitors, and merger control may also need to be updated to reflect the unique features of digital platforms, including additional aspects such as data privacy exploitation and the use of data as a competition variable. Depending on their competition issues and capacity to use antitrust enforcement to address those issues or benefit indirectly from decisions in other jurisdictions, countries might need ex ante regulations to complement antitrust enforcement (Nyman, Ruival, and Begazo 2023).

Some traditional antitrust tools require adaptation when it comes to digital markets. Cases involving digital platforms require competition authorities to tailor their analyses to multisided markets and consider the impact of the control of data in their assessments. This need complicates the definition of markets as well as the assessment of market shares, market power/dominance, and the effect of anticompetitive practices. Most common frameworks for defining markets—which typically rely on price, such as the hypothetical monopolist test—can be more difficult to apply to digital platforms that provide nominally free products. Therefore, competition analysis in digital markets may need to (1) broaden the concept of consumer welfare beyond prices, and (2) consider new dimensions of competition, such as personal data protection. In addition, platforms typically exist in a digital ecosystem in which providers of complementary digital products interconnect and regularly exchange data to provide consumer products. Thus, competition authorities need to understand how competition restrictions affect these complementary products as well as the direct effects on the users of a platform.

Addressing these often-complex cases will require bolstering state capacity and government institutions. First coordination among a diverse set of institutions (for example, competition, data protection, and consumer protection authorities, and sectoral regulators) is needed to use the best available instruments to address risks. Given the dynamic nature of digital technologies and business models, agile regulation necessitates frequent interaction and knowledge exchange between public institutions and the private sector. Coordination among local institutions is relevant, particularly in cases involving large international players, when coordination between countries will play a key role in developing appropriate frameworks—as in Grab's acquisition of Uber's operations in eight countries (Cambodia, Indonesia, Malaysia, Myanmar, the Philippines, Singapore, Thailand, and Viet Nam). Competition authorities in the Philippines and Singapore collaborated to undertake a full analysis of the merger, finding both risks to prices and barriers to entry. Singapore decided to impose structural remedies whereas the Philippines did not (Grab 2018).

Finally, countries can use noncompetition-specific regulation and nonregulatory means to support contestability of markets. Data access and exchange initiatives can enable competition—for example, in the financial sector open banking facilitates the growth of financial technology (fintech), some of which is based on digital platform (multisided markets) business models. Nonregulatory interventions include support for local digital ecosystems (to prevent markets from tipping) and competition-neutral government platforms to support interoperability (for example, for payments and digital identification).

Fintech
Policy makers and financial regulators are eager to foster the benefits of digital transformation but are also mindful of various regulator challenges that emerge. Fintech can promote poverty alleviation and economic growth by enhancing financial efficiency, inclusion, competition, and innovation; but decision-makers must weigh these benefits carefully against challenges and risks associated with financial stability and integrity; cyber and operational risks; data, consumer, and investor protection; fair competition; and regulatory arbitrage. Ultimately, all forms of financial services provision may give rise to similar risks—for example, fintech firms might become too big and systemically important, leading to the same financial stability concerns that large banks pose.

Decision-makers need to address concerns about market concentration in a balanced way. In some markets, entry of new large and well-resourced players—for example, platform firms (discussed in the previous section)—can lead to more intense beneficial competition, with positive impacts on price and quality. However, the business models of some new entrants also grant unregulated entities significant control over access to a marketplace or to large troves of personal data, resulting in abuses in some markets. Furthermore, Big Tech crossovers into finance mean that monopoly power that might have been beneficial in one market can now be wielded, potentially detrimentally, in another. China's recent regulatory crackdown on Big Tech firms was motivated, at least partly, by the fact that these firms had grown too big within a short time, providing among others extensive financial services without being properly regulated.

Among the various approaches EAP regulators have taken to regulate fintech activities, three approaches predominate: regulate, wait and see, and test and learn. Overall, we observe a more cautious approach to "first-generation" fintech activities such as mobile payments and e-money, and a more active, hands-on approach to newer fintech business models such as digital lending and equity crowdfunding. The first attempts of EAP regulators to regulate mobile money and extend agent banking mirror, to a certain extent, the approaches African regulators have taken. China's approach to mobile payments can be best characterized as "wait and see." The Philippines took a similar approach to mobile money, allowing mobile network operators Globe and PLDT (through its subsidiary Smart) to pilot new mobile money products to their customers in 2004. These pilots were closely supervised by the central bank, which undertook regulatory action only five years later.

Although the "wait and see" approach has worked for digital payments, credit products have warranted more proactive regulatory interventions. For instance, peer-to-peer lending in China reached significant volumes before the regulatory crackdown began in 2017 and culminated in the November 2019 requirement that all platforms close or convert to regulated small loan providers within two years. Numerous instances of fraud and ponzi schemes (by some accounts up to 40 percent of platforms were problematic) had resulted in customer losses. Oversight was tightened, the industry shrank, and eventually peer-to-peer marketplaces were mandated to convert to regulated small loan providers. When Big Tech firms reached the level of systemic importance (too big to fail) in some economies, notably in China, authorities actively formulated regulatory responses to these risks.

EAP authorities have also started to issue new regulations or adapt their existing regulatory frameworks to changing needs. Open banking regulations represent a proactive attempt by regulators to foster entry of new business models and entrants, and bring these activities within the regulatory perimeter. The key motivation for open banking has been to foster competition and provide a pathway for fintech firms to offer services efficiently instead of having to rely on unreliable and risky processes such as screen scrapping. In the region, most open banking regimes are regulation driven, except in Korea and Singapore whose regimes are market driven. EAP authorities have adapted their existing frameworks and issued specific licensing frameworks for digital-only banks. For example, virtual banks in Hong Kong SAR, China; internet-only banks in Korea and Taiwan, China; and digital banks in Singapore have restrictions on physical presence and a focus on financial inclusion, and must meet the fundamental requirements for banks (that is, anti-money-laundering/combatting the financing of terrorism and consumer protection rules, risk management, and certain prudential requirements like minimum capital).

Bringing these new activities and business models within an authority's regulatory ambit also has an impact on supervisory approaches. Establishing a risk-based framework to prioritize supervisory actions and calibrate supervisory intensity becomes relevant. Further, supervisors will need to marshal new skills through strategic staffing, partnerships, and industry collaborations. Strengthening and expanding data-sharing and collaboration frameworks among domestic authorities and at the international level are important. As the fintech market evolves, ensuring an orderly exit of unviable market players could become critical and necessitate strengthening of wind-down processes and tools and financial sector safeguards.

Finally, in addition to adjusting prudential and supervisory frameworks, the design and governance of financial infrastructure is a key policy lever to fully harness efficiency gains and safeguard competition. Ensuring open, fair, and transparent access to these infrastructures become critical to provide a level playing field and allow new entrants a fair chance to compete with incumbents. Doing so involves, first, efficient and open payment systems. All Association of Southeast Asian Nations (ASEAN) members[5] have either already implemented or are in the process of implementing a fast payment system; thus, supporting their integration would be an effective way of improving cross-border payments. Pilot projects have established specific cross-border QR payment links with Malaysia, Singapore, and Thailand, and those pilots are now being expanded to cover other economies as well. Second, financial services, especially beyond basic payments, require effective digital identification systems. Australia; Hong Kong SAR, China; Indonesia; Japan; Malaysia; Mongolia; the Philippines; Singapore; and Thailand have introduced such systems. The government of Singapore introduced Singpass in 2003 to enable Singaporeans to access government services online. In 2017, Singpass evolved to become part of the country's national digital identification service. The Philippine Statistics Authority launched PhilSys as the country's foundational digital identification system. As of April 2023, PhilSys had more than 78.2 million citizens registered.[6]

Building skills and infrastructure

If they have the relevant skills and access to the appropriate infrastructure, people can take advantage of the new opportunities that emerge with services. A key challenge is determining how far the market and private institutions can be relied on to deliver the skills and infrastructure needed by the digital services economy and what role the state must play in provision, financing, regulation, and other policies to enable markets to operate more efficiently.

Skills

Although strong foundational skills are essential (refer to Afkar et al. 2023), people also need the more sophisticated skills relevant for the new services economy. Therefore, investments and policies to enhance higher education and digital skills are important.

Tertiary education expansion, returns, and mismatches
Enrollment in tertiary education has grown considerably in every economy in the EAP region. Most EAP economies expanded tertiary enrollment faster than other countries once they reached middle-income levels (figure 4.5). Notable exceptions are Cambodia, the Lao People's Democratic Republic, and the Pacific island countries, where enrollment rates remain below what is expected at their income level.

The expansion of tertiary education in the region has relied in part on private sector provision. Although they increased between 2000 and 2021, enrollment rates remain below 50 percent in most countries (figure 4.6, panel a). More than half of tertiary education students attend a private institution in Cambodia, Indonesia, and the Philippines, but less than a quarter do so in most other countries (figure 4.6, panel b). Moreover, this share has increased slowly compared to other regions and has even declined in some countries. It is important to understand what factors inhibit the private sector supply of higher education and what reforms are needed.

The demand for tertiary education graduates kept pace with the supply expansion until the early to mid-2010s. As shown in figure 4.7, average returns to tertiary education remained largely unchanged during the first decade of the twenty-first century in countries like Indonesia, the Philippines, and Thailand, and even increased in

FIGURE 4.5 **Most EAP economies increased tertiary education enrollment faster than other economies once they reached middle-income levels**

Tertiary education enrollment and GDP per capita, EAP economies and comparators, by income level, circa 1970–2022

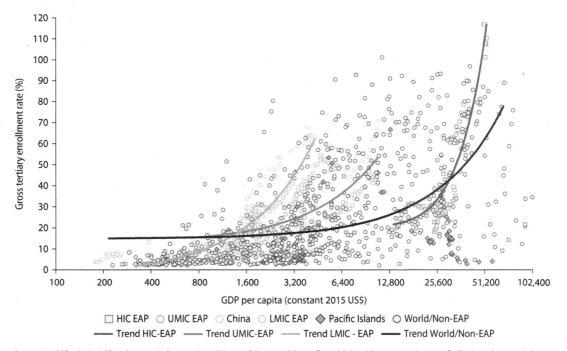

Sources: World Bank, World Development Indicators; United Nations Educational, Scientific and Cultural Organization Institute for Statistics (as recorded in World Development Indicators), accessed November 2023.
Note: The figure displays the evolution of tertiary education enrollment shares and real GDP per capita (in constant 2015 US$) between circa 1970 and circa 2022 for countries with at least three years of data. The classification of lower-middle-, upper-middle-, and high-income countries (LMIC, UMIC, and HIC) uses World Bank economies' income groupings as of the most recent year available. EAP = East Asia and Pacific.

Cambodia, Mongolia, and Viet Nam despite expanded enrollment, albeit from a low base. It takes time for the tertiary supply expansion to significantly affect the actual stock of college-educated workers in the labor force. As cohorts of college graduates enter the labor force, the college earnings premium likely faces increasing downward pressure.

In fact, starting in the mid-2000s, signs show the supply of skills running ahead of the demand in some EAP countries. Recent evidence for China (Hanushek, Wang, and Zhang 2023), Malaysia (World Bank 2024), Thailand (World Bank 2023), and Viet Nam (Banh, Dao, and Glewwe 2024) indicates that the returns to tertiary education started to fall in the mid-2010s. For instance, as shown in figure 4.8, average returns in China and Viet Nam fell significantly (panels b and d) as new college graduates entered the labor force, especially in China (panels a and c).

Other evidence points to a mismatch between skills (capacities) and the demand (opportunities) for high-skilled employment. A rough indicator of the skills mismatch

FIGURE 4.6 **Tertiary education enrollment rates have increased but remain low in most countries, with variations in the share of private institution enrollment**

Tertiary education enrollment, total and private, selected EAP countries

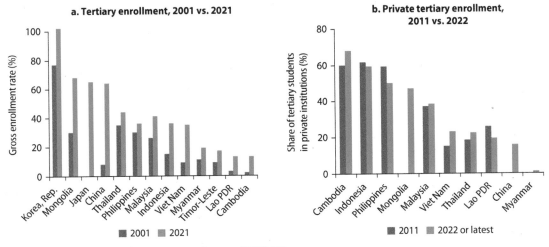

a. Tertiary enrollment, 2001 vs. 2021

b. Private tertiary enrollment, 2011 vs. 2022

■ 2001 ■ 2021

■ 2011 ■ 2022 or latest

Source: International Labour Organization Department of Statistics, ILOSTAT, 2022.
Note: 2011 data unavailable for China and Myanmar.

FIGURE 4.7 **Returns to tertiary education remained stable until the early 2000s despite tertiary enrollment expansion**

Changes in average returns to tertiary education and gross tertiary enrollment, selected EAP countries, circa 2000–2010s

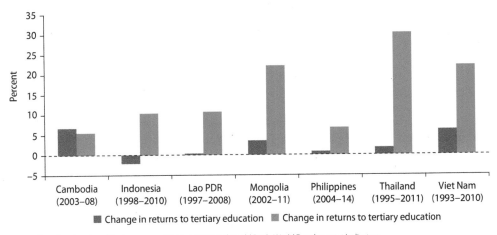

■ Change in returns to tertiary education ■ Change in returns to tertiary education

Sources: Based on data from Montenegro and Patrinos 2023 and World Bank, World Development Indicators.

FIGURE 4.8 **Returns to higher education have declined recently with a surge in supply of graduates**

Share of workers with tertiary education and the returns to tertiary education, China and Viet Nam

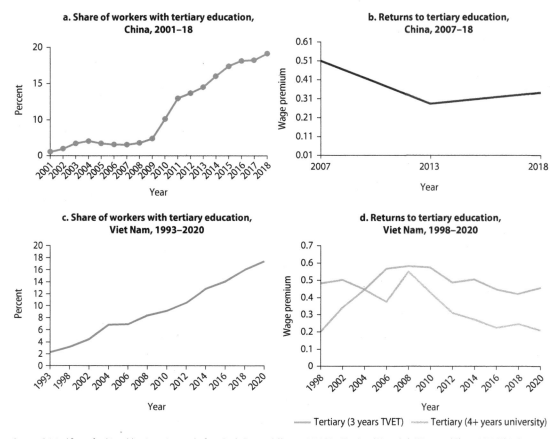

Sources: Original figure for this publication using results from Banh, Dao, and Glewwe 2024 (Viet Nam) and Hanushek, Wang, and Zhang 2023 (China).
Note: TVET = technical and vocational education and training.

is the extent to which workers are employed in occupations that match their education levels. The International Labour Organization uses this measure to estimate the share of overeducated or undereducated workers. For instance, an individual with a college degree working as a clerk would be considered overeducated, whereas an individual with primary education working as a manager would be considered undereducated. Figure 4.9 reports evidence of the prevalence of undereducation in many EAP countries and overeducation in countries like Korea, Mongolia, and Viet Nam.

These figures mask differences in the nature of the misalignment between skills supply and demand. Whereas Mongolia and Viet Nam both have unusually large shares of jobs filled by overeducated workers, the two countries differ crucially in the composition of the overeducated. In Viet Nam, 72 percent of those deemed

FIGURE 4.9   Emerging signs indicate that several EAP countries are struggling to keep up with the pace of demand for high-skilled employment

Share of under- and overeducated workers, selected EAP countries

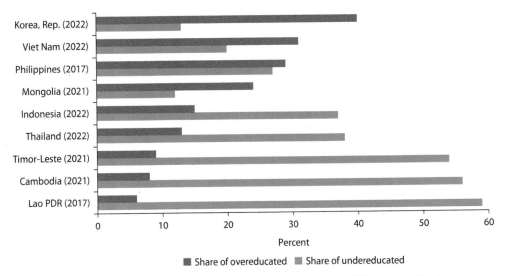

■ Share of overeducated   ■ Share of undereducated

Source: Original figure for this publication using data from the International Labour Organization (ILO) Department of Statistics, ILOSTAT, 2022.
Note: The ILO matches occupation types with education levels to establish whether workers are overeducated or undereducated. For instance, an individual with a doctoral degree working as a clerk would be considered overeducated, whereas an individual with primary education working as a manager, would be considered undereducated. EAP = East Asia and Pacific.

overeducated work in an "elementary occupation" and have either primary or secondary education. By contrast, this figure is only 30 percent for Mongolia; instead, in Mongolia, 66 percent of those deemed overeducated hold tertiary degrees and work in occupations that often require only a secondary education. The data suggest that Mongolia's supply of tertiary graduates may have outpaced the demand for such skills. In Viet Nam, the prevalence of both overeducation and undereducation might reflect the success of the education system in ensuring nearly universal quality basic education, and a labor demand that has not expanded at a sufficient pace to create opportunities in more skill-intensive activities.

A key aspect of the skills mismatch is the growing need for digital and behavioral skills in both services and advanced manufacturing. Data from the International Telecommunication Union, presented in figure 4.10, reveal how developing EAP countries lag developed countries like Japan, Korea, and Singapore in terms of both basic digital skills (such as sending emails or processing a document) and more advanced digital skills (such as using a spreadsheet or coding).

Furthermore, as shown in panel b of figure 4.10, an age divide exists in the countries with available data: digital skills are much scarcer among the adult population (ages 25–74) than among youth (ages 15–24). There is also evidence that, along with digital skills, jobs in modern services require workers to have strong social skills for tasks that are less susceptible to automation (Cunningham et al. 2021). Countries need not only to equip future workers with the skills that will be highly demanded

FIGURE 4.10   Developing EAP countries have low levels of digital skills (basic and advanced) compared to developed Asian economies

ICT skills, by type of skill and age group, EAP countries and comparators

a. ICT skills, 2021 (or latest available)

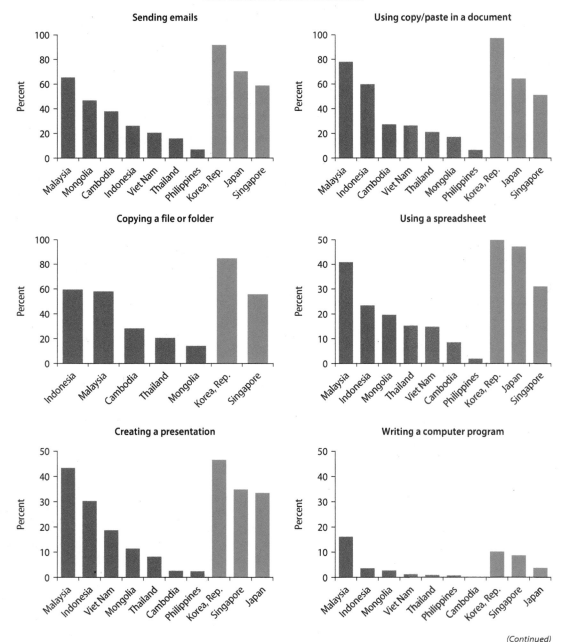

(Continued)

FIGURE 4.10 **Developing EAP countries have low levels of digital skills (basic and advanced) compared to developed Asian economies** *(Continued)*

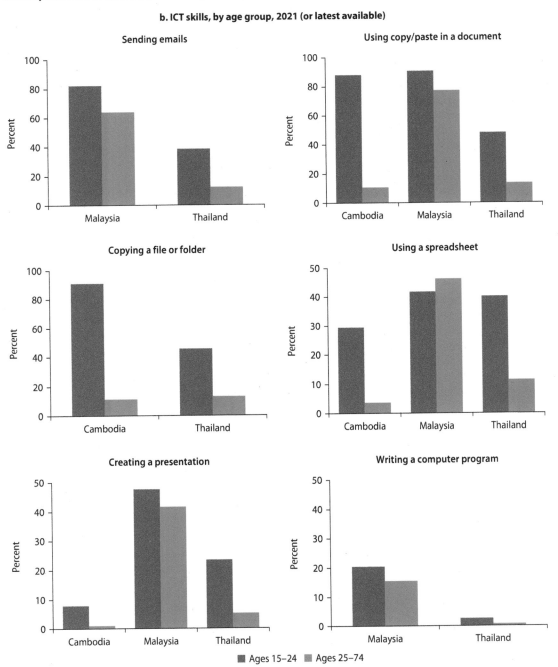

b. ICT skills, by age group, 2021 (or latest available)

Source: International Telecommunication Union.
Note: The proportion of youth and adults with ICT skills, by type of skill, is defined as the percentage of individuals who have undertaken certain ICT-related activities in the last three months. Last available years: Cambodia (2017), Indonesia (2017), Japan (2021), Malaysia (2021), Mongolia (2021), Philippines (2019), Republic of Korea (2021), Singapore (2021), Thailand (2019), and Viet Nam (2021). EAP = East Asia and Pacific; ICT = information and communication technology.

tomorrow but also to equip current workers with the skills demanded today. To this end, policies related to governance, regulation, information, and coordination will play an important role in ensuring the alignment of the supply of tertiary education to the dynamic evolution of skills demand in EAP.

Policy reforms to meet the skills needs in EAP countries
A robust tertiary education system needs to serve the needs of students and economies today and in the future by offering equitable access to relevant content of the requisite quality. Achieving these goals requires all agents—government, the private sector, students, and their families—to play a role. Governments need to ensure equitable access through direct provision, financing, or other targeted policies to remedy market failures and align supply to demand. The private sector can be a key partner in the provision of program offerings aligned with the economy's needs. Students and families should make informed choices to invest in programs of study based on their interests, abilities, academic readiness, and the expected returns to their investments.

Whereas basic education is largely provided by the public sector in most countries (due to equity concerns and the strong social externalities), tertiary education systems around the world are a mix of public and private institutions. Examples of high-performing systems exist in both cases: the United States has predominantly public enrollment (over 70 percent) but has both public and private universities that top world rankings. Private provision dominates in countries like Japan, Korea, Indonesia, and the Philippines, whereas public provision is larger in China, Singapore, and Thailand. In all countries, the growing demand for tertiary education has been met with an increasing role for private higher education institutions.

Regardless of the mix of public and private provision in higher education, a strong tertiary education system provides diversified pathways to skills development that give students choices and meet the economy's needs. Students differ in their interests, academic readiness, and talents; and the economy needs various types of skills—from engineers, doctors, and managers to nurses, operators, and technicians. Whereas some pathways require four or five years of coursework, shorter programs may offer a better fit for some students. To serve these needs, a range of postsecondary institutions—universities, community colleges, polytechnic institutions, and technical training institutions—need to come together into one single well-functioning ecosystem.

In designing policies and regulations to steer and govern this ecosystem, policy makers need to recognize the risks of government failures even as they seek to remedy market distortions in the form of credit constraints, information asymmetries, imperfect competition, and coordination failures. Each of these distortions calls for a different set of policies, and together they might even constitute a case for more direct government involvement:[7]

- *Credit constraints* may call either for targeted government subsidies or for other funding mechanisms that address credit market imperfections.
- *Information asymmetries* call for information provision and regulation to protect the interests of students.

- *Imperfect competition* calls for regulatory oversight to prevent anticompetitive behavior and ensure an ecosystem that gives students choices that fit their interests and talent, and value for money for their investments.
- *Coordination failures* call for public-private partnerships to ensure adequate investments in skills, including through targeted subsidies, guarantees, or direct public provision for certain types of higher education that involve externalities or have nascent demand.

Credit constraints

Credit constraints may prevent students, particularly those from low-income families, from pursuing tertiary education. Credit markets typically are not well suited to finance tertiary education investments because of the lack of collateral or guarantees and the uncertainty of realizing the long-term returns to higher education. Thus, credit markets tend to provide less finance to tertiary education than is socially optimal. There is a role for funding mechanisms that enable expansion and equality of access while incentivizing informed choices, student effort, and desirable economic outcomes.

Governments in EAP typically use a combination of total or partial tuition subsidies, loans, and scholarships to address credit constraints. Overreliance on "free" or subsidized tax payer financing can disproportionately benefit wealthier students (especially in public education systems of low quality) or lead to inadequate per-student spending that undermines quality. Several EAP countries—chiefly China, Korea, Malaysia, Mongolia, Thailand, and Viet Nam—rely on student loans to complement public finance while creating more powerful incentives for students to internalize the cost and benefits of different programs and to encourage completion of their education. Despite having improved access to tertiary education, these loan schemes often tend to benefit students from more affluent backgrounds and result in high debt burdens for graduates of low-income families by shifting all the uncertainty arising from future employability to individuals and their families (Barr 2014; Chapman and Dearden 2022; Di Gropello 2012).

Numerous countries—including Japan, Korea, New Zealand, the United Kingdom, and the United States—have adopted income-contingent loans, pioneered in Australia in 1989, because these loans insure against the risk of future low incomes by making repayments contingent on future earnings above a given threshold. These schemes vary in terms of their design, eligibility, interest rates, and debt forgiveness, but overall have resulted in lower repayment burdens and default rates when the required functional tax collection and personal identification systems are in place (for example, employer withholding of personal income taxes) (Chapman and Dearden 2022). These schemes, combined with merit- and need-based financial aid programs conditional on good academic standing, can deliver a better balance between expanding enrollment and better student choices and outcomes than universal subsidies, a feature of public systems in some EAP countries (Chapman and Dearden 2022).

Information asymmetries

The market for tertiary education is characterized by strong information asymmetries. Students and their families are expected to undertake a cost-benefit analysis to decide on whether to enroll in tertiary education, what course of study to pursue, and which institution. They must weigh the costs—tuition, other education expenses, and the opportunity cost of a salary that can be earned with a high school diploma—against the uncertain stream of future earnings of tertiary education. They often make these choices with little information on the quality and variety of institutions and program offerings or on the labor market prospects (uncertain returns) of different programs of study. Decision-making barriers (for example, cognitive biases) may lead students to enroll in programs of low quality or that do not fit their talents.

Numerous countries (as diverse as Australia, Bangladesh, Canada, Chile, Colombia, most European countries, and the United States) regularly collect and provide information on the outcomes of tertiary education graduates using graduate tracer surveys and administrative (social security) data. For many years, countries like Chile and Colombia have used a combination of administrative and survey data to track and provide information on graduates' employability and job vacancies. Germany requires a system to regularly trace graduates as part of the accreditation of tertiary education institutions, and the United Kingdom mandates that, in order to access public funds, universities must submit data on graduates' labor market performance. Except in Korea, Malaysia, and Singapore, use of these mechanisms is not yet well established in the EAP region. Countries like Indonesia, Thailand, and Viet Nam have started to carry out graduate tracer studies; but most EAP countries do not yet regularly use administrative, labor force, and enterprise data to provide timely and accessible information on the outcomes of various types of tertiary education to inform student decisions and ensure relevance to the labor market.

Imperfect competition
Imperfect competition pervades tertiary education markets. Establishing a high-quality tertiary education institution may involve high costs that can limit the entry of providers and concentrate the market, especially with inadequate (too lax or burdensome) regulations on the entry and exit of institutions. Institutions purport to offer a differentiated service—whether real or through branding—and leverage it to gain market power. For public institutions, tension can exist between autonomy and accountability, and their offerings of free or highly subsidized tuition give them an advantage relative to private providers.

Sound regulations need to (1) provide clarity for private sector participation; (2) strike an adequate balance between ensuring minimum quality standards and consumer protection and restraining the entry (exit) of good-quality (low-quality) institutions;[8] (3) distinguish between for-profit and not-for-profit providers, and the different purposes and market segments served; and (4) create mechanisms to allow students to move between institutions and across academic and technical tracks. An independent quality assurance agency can oversee compliance with quality standards and the accreditation of institutions and programs. These standards should focus on outcomes (such as data on student performance and graduates' employability) or inputs that drive quality. Well-functioning governing boards or university councils can allow the government to focus on stewardship of the overall system rather than on directing how public universities should be run. Performance-based financing can be used to nudge accreditation, to tie public funding to outcome-oriented performance measures (such as graduates' employability and graduation rates in certain fields) or to tuition subsidies for students so they can exercise choice over public and private providers.

Several EAP countries carried out many such reforms starting in the 1990s—Korea, followed by Malaysia, Indonesia, and Thailand, and then a second wave of reforms in the 2000s (World Bank 2011). Most EAP countries have established quality-assurance agencies as independent bodies; however, in many countries, accreditation remains voluntary and does not cover all public and private tertiary institutions. A few EAP countries have started to gradually use performance-based funding. In 2020, Indonesia introduced performance-based incentives for public universities tied to the achievement of a few performance indicators, including closer cooperation with industry partners.

The experiences of Japan, Korea, and Singapore and recent reforms in China and Viet Nam offer useful lessons on how to foster an enabling environment for private provision and to ensure quality and variety in supply to meet the growing and differentiated demand for tertiary education. Several countries in Latin American and the Caribbean—notably Chile—have used performance-based funding with some success to nudge accreditation and to foster quality improvement. In the United States, the California Master Plan for Higher Education is a leading example of a system that provides a diversity of program offerings. It comprises institutions that serve students with an academic and research orientation, students pursuing four- to five-year degrees with less research orientation, and community colleges that serve students pursuing shorter programs in practical fields of study.

Coordination failures
Finally, coordination failures can also interfere with the higher education system's ability to form the skills required by economies today and in the future. Expanding tertiary education offerings in science, technology, engineering, and mathematics fields is particularly important to ensure the number of engineers and research scientists an economy would require for engaging in more sophisticated economic activities. For instance, an economy may suffer a shortage of computer programmers and data scientists needed to attract investments in modern services because today's market wages reflect current demand conditions for these skills rather than potential future demand. Moreover, because of the lag in response of the supply of tertiary graduates to the labor market demand for more skills, countries face a real risk of coordination failures: without a supply of workers with the right skills, private firms refrain from investing in high-potential sectors, including modern services; at the same time, without job opportunities, workers do not invest in these skills. That is, inadequate skills and insufficient jobs can lead to a vicious circle of low skills and insufficient investment in the activities that can help countries move to produce and trade higher-value-added manufacturing and services.

Several of today's high-income economies in EAP have become global examples of producing the skills required to fuel economic growth, and their experiences provide valuable examples of success (Almeida, Behrman, and Robalino 2012; Ashton et al. 2002). In the 1960s and 1970s, Korea, Singapore, and Taiwan, China, overcame the coordination failures common to low- and middle-income countries by effectively aligning the incentives of households, employers, and training providers. Governments in all three places pursued strategic plans that created a demand for highly skilled labor, established coordination mechanisms such as skills sector councils, and subsidized training to meet this demand. A ready supply of high-skilled labor subsequently spurred the growth of the technology-intensive sectors that are the engine of these economies today. The government of Korea has arranged partnerships between private employers (including global players like Hyundai Motor Company and Samsung Electronics), universities, and the Meister Vocational Schools to support curriculum development, create industry-education adjunct teachers, and provide students with workplace training and employment opportunities after graduation. Singapore's SkillsFuture program supports continuous learning and skills upgrading based on competencies acquisition and incentivizes universities and training institutions to work closely with employers to offer programs that meet the evolving needs of the economy.

Digital infrastructure

Digital connectivity, especially through mobile technology, has expanded dramatically since 2010. Today, the EAP region is home to 1.5 billion active internet users, accounting for over 70 percent of the region's total population (figure 4.11, panel a).[9]

Although mobile broadband subscriptions have expanded rapidly, increasing from 437 million in 2013 to 2.0 billion in 2021, only 28.2 percent of the EAP population has access to fixed broadband subscriptions, less than in Europe and Central Asia, Latin America and the Caribbean, and North America, but well above

FIGURE 4.11   Internet use has expanded rapidly in the EAP region

Share of population using the internet and share of internet poor, EAP and comparators

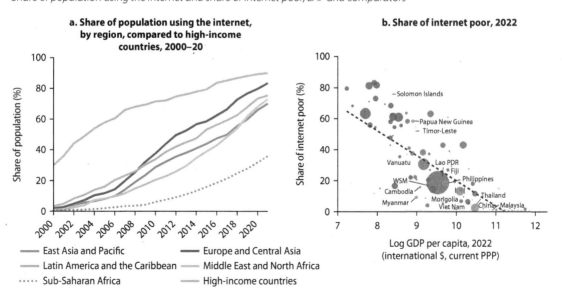

Sources: International Telecommunication Union; World Data Lab.
Note: In panel b, "internet poor" shows the share of people unable to afford a minimum package of mobile internet (that is, 1 gigabyte per month, 10 megabits per second download speed) for 10 percent of total individual spending. The size of the bubbles represents the share of the population in each country who are internet poor. EAP = East Asia and Pacific; PPP = purchasing power parity.

levels in the Middle East and North Africa, South Asia, and Sub-Saharan Africa. Moreover, data on the cost of different bundles of broadband services suggest that fixed-line broadband services tend to cost much more in EAP, in many countries exceeding 2 percent of average national income, the target set by the Broadband Commission.

Despite progress in expanding digital connectivity, gaps in access remain—both within and across countries. According to data from the International Telecommunication Union, over a quarter of the population in EAP remains unconnected in 2021, with poorer households more likely to suffer from limited access. Those data show that the poorest 20 percent of households in EAP are 43 percent less likely to access the internet than the richest 20 percent. In total, 111.3 million people—5.3 percent of the total EAP population—live in "internet poverty," meaning they cannot afford access to a basic minimum package of mobile internet access (1 gigabyte per month, 10 megabits per second download speed) for 10 percent of total individual spending (figure 4.11, panel b). Small island nations in the Pacific have the most pronounced access gaps and internet poverty; however, even in Cambodia, Indonesia, Lao PDR, and the Philippines, 15–20 percent of the population currently cannot afford a basic mobile internet plan. By contrast, more than 97 percent of the population in China, Malaysia, Thailand, and Viet Nam can afford a minimum package of mobile internet access.

Regional differences in infrastructure development and connectivity often exist within countries. For example, Indonesia has large infrastructure gaps between its western and eastern parts, leading to a higher concentration of internet users in Java, Kalimantan, and Sumatra, and a lower proportion in Maluku, Nusa Tenggara, and Papua (World Bank 2021a). The digital divide is particularly acute for the underserved population. As of January 2022, more than 12,500 villages and 104,000 schools across Indonesia still have no internet access (Amanta 2022).

In addition, limited network speed remains a constraint. Download speed in EAP averages only 40 percent of the average in high-income economies, again with stark differences within and across countries. Despite high mobile cellular coverage, the high-speed mobile coverage rate (that is, population covered by at least a fourth-generation (4G) mobile network) averaged only about 65 percent across economies, lower than the average of 69 percent in low- and middle-income economies. Stark differences exist across countries between early adopters and slow starters as mobile technology evolved from 3G to 5G. In the case of 4G, countries like Malaysia and Thailand reached penetration levels comparable to the most advanced economies, with the rest of ASEAN performing between the levels of South Asia and Western Europe, and others like the Pacific islands remaining at the level of Sub-Saharan Africa or lower.

Moreover, the relatively low penetration of fixed internet technologies constrains access to high-speed internet. Because of the quality of both mobile and fixed infrastructure, wide gaps in internet coverage, quality (measured by download speed), and cost (measured as basket price as percent of gross national income per capita) exist across countries (table 4.1). For example, according to the Ookla Speedtest Global Index in 2020, mobile internet and fixed broadband internet speeds in the Philippines were more than 55 percent slower than global averages for these services. Consequently, the Philippines ranked 111 out of 139 countries for mobile internet speed and 107 out of 176 countries for fixed broadband speed (Gómez 2021).

TABLE 4.1   Fixed and mobile broadband internet service varies widely across EAP countries

Access, quality, and cost of fixed and mobile broadband, EAP countries

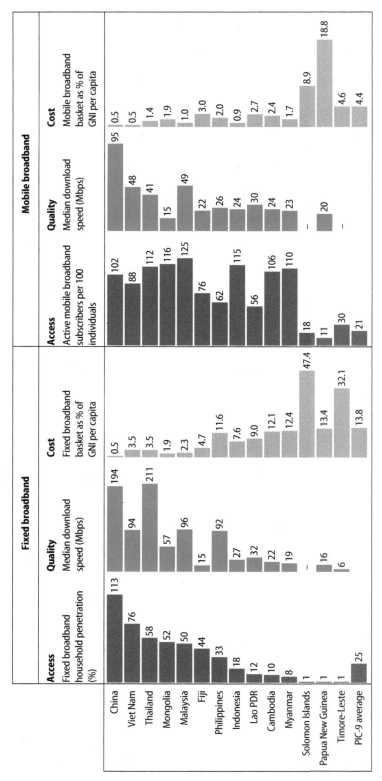

	Fixed broadband			Mobile broadband		
	Access Fixed broadband household penetration (%)	**Quality** Median download speed (Mbps)	**Cost** Fixed broadband basket as % of GNI per capita	**Access** Active mobile broadband subscribers per 100 individuals	**Quality** Median download speed (Mbps)	**Cost** Mobile broadband basket as % of GNI per capita
China	113	194	0.5	102	95	0.5
Viet Nam	76	94	3.5	88	48	0.5
Thailand	58	211	3.5	112	41	1.4
Mongolia	52	57	1.9	116	15	1.9
Malaysia	50	96	2.3	125	49	1.0
Fiji	44	15	4.7	76	22	3.0
Philippines	33	92	11.6	62	26	2.0
Indonesia	18	27	7.6	115	24	0.9
Lao PDR	12	32	9.0	56	30	2.7
Cambodia	10	22	12.1	106	24	2.4
Myanmar	8	19	12.4	110	23	1.7
Solomon Islands	1	—	47.4	18	—	8.9
Papua New Guinea	1	16	13.4	11	20	18.8
Timor-Leste	1	6	32.1	30	—	4.6
PIC-9 average	25	—	13.8	21	—	4.4

Sources: International Telecommunication Union; Ookla; TeleGeography.

Note: Because not all Pacific island countries had data available, the table shows numerical average of the available data points for PIC-9. GNI = gross national income; Mbps = megabits per second; PIC-9 = Kiribati, Marshall Islands, Micronesia, Nauru, Palau, Samoa, Tonga, Tuvalu, and Vanuatu; — = not available.

Map 4.1 demonstrates the region's progress in high-speed connectivity between the end of 2019 and the beginning of 2023. Malaysia, the Philippines, and Viet Nam registered a strong increase in the speed of the connections. The map also shows the large within-country heterogeneity in internet coverage and speed, with better internet quality found mostly in urban and larger cities, especially for fixed-line broadband internet connection.

These physical access and infrastructure gaps amplify other inequities in terms of digital skills and together dampen the potential to reap digital dividends. The relatively low penetration of high-speed mobile (4G/5G) and broadband fixed lines reduces the ability to run the most advanced digital applications—for example, cloud computing—especially in business-to-business services, a key driver to higher-productivity growth. The limited access in poorer countries and, within countries, among lower-income households constrains the potential of poorer people and communities to benefit from increased access to public services and economic opportunities. Therefore, it is important to promote investment in expanding and upgrading those networks. Doing so will require a multifaceted policy response and the involvement of both the public and the private sector.

First, the regulatory environment is crucial for attracting private investment into network infrastructure. There are legitimate reasons to regulate telecommunications services, including overcrowding of radio frequencies, network and data security, and consumer protection; ensuring that regulations do not create undue barriers to market entry and competition is key. In general, without liberalization of the telecom sector and if barriers to entry exist, incumbent companies—which in several countries remain state-owned, vertically integrated telecommunication companies—have less incentive to expand and preserve the quality of the infrastructure. That is, they can squeeze their captive consumer base without fear of losing customers to alternative providers, often leading to high prices and low-quality services for consumers. The possible presence of intrinsic weaknesses in the country's business climate, which expose investors to risks of appropriation (for example, through excessive taxation), could exacerbate the negative effects of such reduced competition and contestability.

Second, irrespective of such institutional constraints, network coverage might still be limited: reaching less densely populated or intrinsically high-cost areas (like remote islands) might not make business sense because costs exceed expected revenues. In this case, governments have a rationale to subsidize such investments in last-mile connectivity. They could invoke social cohesion reasons (allowing all the population to enjoy a minimum level of digital services) and economic efficiency motivations (internalizing the positive network externalities associated with the expansion of telecom services, or de-risking investments in areas that are particularly prone to natural disasters). Interventions could take the form of the creation of Universal Service Funds, replenished either through public finances or with special contributions on telecom bills.

Finally, constraints arising from the demand side are also important in many countries. First, when large fractions of the population are poor, affordability is a major concern. The policy solution to the affordability problem would be to provide financial support so the poorest segments of society can access digital technologies. Additional demand-side constraints could arise because users lack a sufficient level of digital skills and consider little of the content useful. The first problem requires

MAP 4.1 **Internet connection speed within and across EAP countries, 2019 vs. 2023**

a. Broadband fixed line internet speed, 2019 Q4

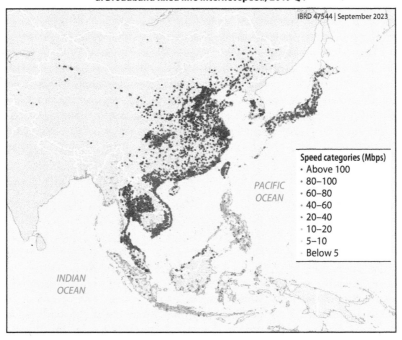

b. Broadband fixed line internet speed, 2023 Q2

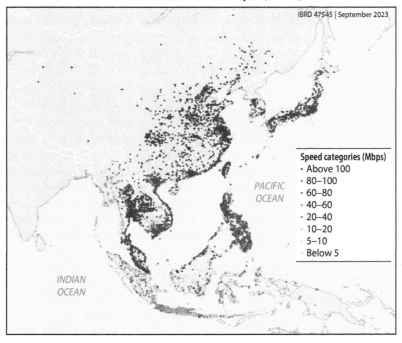

(continued)

MAP 4.1 **Internet connection speed within and across EAP countries, 2019 vs. 2023** *(Continued)*

c. Broadband mobile line internet speed, 2019 Q4

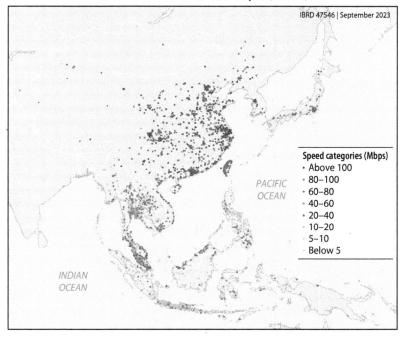

IBRD 47546 | September 2023

PACIFIC OCEAN

INDIAN OCEAN

Speed categories (Mbps)
- Above 100
- 80–100
- 60–80
- 40–60
- 20–40
- 10–20
- 5–10
- Below 5

d. Broadband mobile line internet speed, 2023 Q2

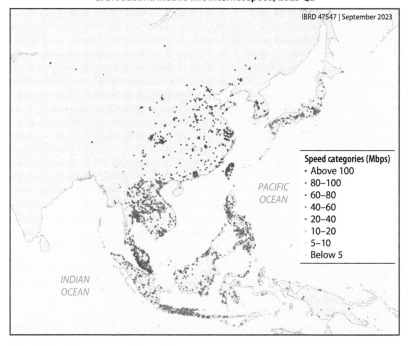

IBRD 47547 | September 2023

PACIFIC OCEAN

INDIAN OCEAN

Speed categories (Mbps)
- Above 100
- 80–100
- 60–80
- 40–60
- 20–40
- 10–20
- 5–10
- Below 5

Source: Original map for this publication using data from Ookla.

governments to invest in training and education. Digital skills are becoming essential for learning recovery and productivity growth, but significant skills gaps persist, particularly in developing EAP countries, as seen in the previous section. The second problem could be mitigated by subsidizing developers that produce apps relevant for the context (for example, those that rely less on text in countries where lack of literacy is a major constraint).

International cooperation on liberalization and regulation

The digital services economy provides both the scope and the necessity for international cooperation. Such cooperation can lead, first, to reciprocal liberalization. Trade negotiations promote greater mutual openness by mobilizing exportable sector interests that seek improved access to foreign markets, to countervail the power of importable sector interests that seek continued protection at home. Second, cooperation on regulation can help address international market failures, such as those attributable to asymmetric information or environmental externalities. In effect, the well-being of consumers in each country can be enhanced if regulators in other countries protect them from the adverse actions of suppliers abroad. The provision of regulatory reassurance from trading partners can provide an additional impulse to liberalization by eliminating the need for precautionary restrictions, for example, on international data flows.

The following subsections discuss how international cooperation on services has not led to much liberalization but has made policies more predictable by legally locking in existing openness. At the same time, regulatory cooperation is beginning to produce benefits in at least two areas: cross-border data flows and policy action to reduce carbon emissions from cross-border transportation services.

Services trade liberalization

Recently, multilateral cooperation at the World Trade Organization has stalled, but economies in the EAP region have participated in two major regional trade agreements.

The Regional Comprehensive Economic Partnership (RCEP), a trade agreement signed by 15 EAP nations in November 2020, covers almost one-third of the global population and global economic activity. It aims to boost growth by lowering barriers to trade and investment in goods and services among member countries. The agreement consolidates and updates existing free trade agreements between ASEAN and its partners, including major economies like China, Japan, and Korea.

The Comprehensive and Progressive Agreement for Trans-Pacific Partnership (CPTPP), signed by 11 countries in the EAP region (some of which are also signatories of the RCEP), aims to promote economic integration among member countries. The agreement covers a wide range of areas, including trade in goods, services, investment, intellectual property, and government procurement. CPTPP eliminates or reduces tariffs on various products, facilitates cross-border trade and data flows, and establishes rules and regulations to enhance transparency and predictability in trade relations.

This section presents preliminary evidence from a newly computed Preferential Services Trade Restrictiveness Index (PSTRI), jointly developed by the World Bank and the World Trade Organizaation. The PSTRI is an index of restrictiveness similar to the STRI presented earlier and based on the text of the commitments of different countries to services trade liberalization under RCEP and CPTPP. The new computation allows for comparison between the restrictiveness of the currently applied regime and the liberalization promised under these new agreements.

In most instances, the values of the PSTRI computed from the texts of RCEP and CPTPP are higher than the scores obtained considering the applied measures. Figure 4.12 reports the results for six services sectors, obtained by comparing the overall STRI computed using applied policies to the overall PSTRI, computed using commitments under RCEP and CPTPP texts. Figure 4.12 lends empirical support to the notion that these recent important preferential agreements will not substantially liberalize trade in services.

However, figure 4.12 reports overall measures for relatively large sectors. Moving deeper, and analyzing PSTRI scores for services subsectors and different modes of delivery, allows for more nuanced conclusions. RCEP and CPTPP present, respectively, 935 and 781 country-subsector-mode combinations for which an applied STRI is also available. Figure 4.13 plots the densities across all these combinations of country-subsector-mode of the differences between RCEP and applied STRI (solid line) and the difference between CPTPP and applied STRI (dashed line). A positive number indicates that the RCEP or CPTPP regime is more restrictive than the actual applied policies.

Three observations stand out. First, both distributions peak at zero, meaning that, on average, both RCEP and CPTPP scores are close to the scores obtained using currently applied policies. Second, both distributions are right skewed, meaning that, for both RCEP and CPTPP, there are more instances in which the scores are higher than under currently applied policies than vice versa. Third, this tendency is more pronounced for RCEP scores than for CPTPP scores, with two notable exceptions. China's STRI scores in computer, communication, and financial services are lower under RCEP commitments; and Malaysia registers lower restrictiveness scores under CPTPP than under either the RCEP commitments or the applied regimes.

These results lead to the conclusion that, even if these agreements have the virtue of credibly locking in existing policy regimes, one cannot reasonably expect to see much more liberalization in services trade come simply from the implementation of CPTPP and RCEP. The main liberalization impetus will have to come from unilateral national efforts, such as those seen recently in the Philippines.

Cross-border data flows

The rapid digitalization of everyday life has ushered in a new kind of globalization, with cross-border data flows emerging alongside traditional flows of goods, services, and capital as a new pillar of the global economy. The ability to use, share, and access data across borders is instrumental in stimulating innovation, enabling data-driven products and services, and fueling economic growth and ideas.

To reap the full benefits of digital trade and commerce while upholding and strengthening data security and privacy, governments need to foster trust in digital transactions both domestically and across borders. One way to support this trust is to improve regulatory predictability for both those who process personal data and

FIGURE 4.12   In most cases, PSTRI scores for the commitment to services trade liberalization under RCEP and CPTPP are higher (less liberalized) than the score computed from applied regimes

Services trade restrictiveness under current and proposed regimes, by sector, selected countries

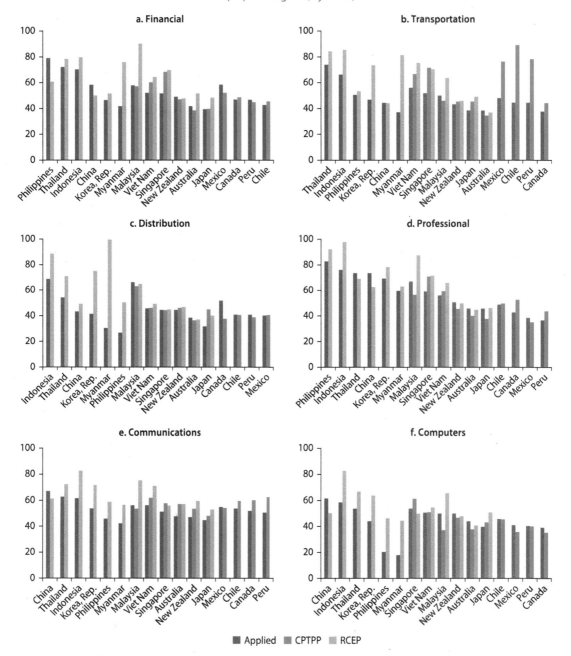

Source: World Bank and World Trade Organization estimates based on World Bank–World Trade Organization Preferential Services Trade Restrictiveness Index methodology.
Note: CPTPP = Comprehensive and Progressive Agreement for Trans-Pacific Partnership; PSTRI = Preferential Services Trade Restrictiveness Index; RCEP = Regional Comprehensive Economic Partnership.

FIGURE 4.13   **Distribution of the differences between PSTRI under RCEP and CPTPP and STRI computed using applied regimes are right skewed, more so for RCEP**

Difference between PSTRI (under RCEP and CPTPP) and applied STRI

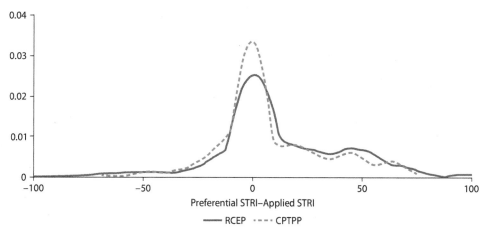

Source: World Bank and World Trade Organization estimates based on World Bank–World Trade Organization Preferential Services Trade Restrictiveness Index (PSTRI) methodology and STRI Database.
Note: CPTPP = Comprehensive and Progressive Agreement for Trans-Pacific Partnership; PSTRI = Preferential Services Trade Restrictiveness Index; RCEP = Regional Comprehensive Economic Partnership.

individuals providing their data. Developing common approaches to enable cross-border data flows of a broad variety of personal and nonpersonal data with the appropriate safeguards while minimizing the compliance burdens can help support these objectives.

World Development Report 2020: Data for Better Lives identifies three regulatory approaches to cross-border data flows: open transfer, conditional transfer, and limited transfer (World Bank 2021b). Applying this framework to the regulatory landscape governing cross-border data flows in EAP shows that countries in the region adopt all three approaches. The Philippines and Singapore fall within the "open transfer" approach, which permits data transfers in principle but requires companies transferring personal data to overseas recipients to handle personal information in a manner consistent with the requirements of local laws. Indonesia, Malaysia, and Thailand take the "conditional transfer" approach, in which legal provisions are formulated as general prohibitions, subject to a list of exceptions. These exceptions primarily include obtaining the individual's consent to the transfer ("consent-first" regimes) or sending data to jurisdictions with an "adequate" or "comparable" level of protection. Viet Nam falls within the "limited transfer" approach and imposes stringent controls over the processing and transfer of personal data, including requiring government inspection and restricting access to data to protect national security and public order (map 4.2).

Although the principles underpinning data protection laws in the region are generally consistent across jurisdictions, implementing regulations and practices vary substantially. The interoperability of regulatory regimes across the region is further affected by the diversity of regulatory implementation mechanisms, which include the Asia-Pacific Economic Cooperation Cross-Border Privacy Rules

MAP 4.2 **Regulatory approaches in selected EAP countries**

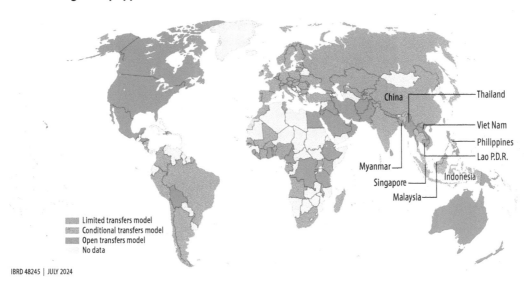

IBRD 48245 | JULY 2024

Source: World Bank 2021b, updated for EAP-6 countries in June 2023.
Note: EAP = East Asia and Pacific.

(CBPRs) certification schemes, standard contractual agreements, and technical and organizational measures (data sandboxes and dataspaces). A key challenge is ensuring that regulators facilitate compliance with legal requirements by developing clear guidelines and other tools that parties conducting a cross-border data transaction can rely on.

Given the divergence in regulatory approaches and implementation practices, there is no one-size-fits-all solution, but the following broad reform directions could guide action across the region (refer also to box 4.1).

First, foster institutional collaboration to *support a harmonized governance framework:*

- Engage in multistakeholder cooperation through participation in intergovernmental (Group of Twenty); multilateral (Organisation for Economic Co-operation and Development, World Bank, and World Trade Organization); regional (ASEAN, Council of Europe Data Protection Convention 108+); and bilateral and multilateral trade forums (EU–UK Trade and Cooperation Agreement, and the proposed ASEAN Digital Economy Framework Agreement).
- Strengthen institutional capacity by providing data protection regulators with the resources needed to upgrade their skills and capacity to address regulatory bottlenecks, improve coordination, and provide greater guidance to businesses.
- Establish and expand on regional platforms for better coordination and cooperation among regulators across countries, promote the sharing of knowledge and best practices, improve coordination of enforcement efforts, and achieve more effective responses to cross-border data breaches. The ASEAN Data Protection and Privacy Forum is one such platform.

Box 4.1. Global initiatives to optimize cross-border data flows

The international policy community has developed several initiatives to facilitate a more interoperable regime for enabling cross-border data flows by establishing a "trust framework" for governing the secure processing and sharing of personal data, especially across borders.

At the multilateral level, the Group of Seven and Group of Twenty are developing common strategies, principles, and norms aimed at providing an overarching framework that countries can distill into national policy and regulatory frameworks. Such initiatives have been promoted through regional organizations (the Association of Southeast Asian Nations and the African Union), bilateral trade agreements (the EU–UK Trade and Cooperation Agreement and the proposed Association of Southeast Asian Nations Digital Economy Framework Agreement), and conventions (such as the Council of Europe Data Protection Conventions 108 and 108+).

At the plurilateral and bilateral levels, several preferential trade and digital economy agreements also address cross-border data flows and trust. These agreements include nonbinding guidance on data flows, with broad provisions affirming the importance of working to maintain cross-border data flows (examples include the Republic of Korea–Peru Free Trade Agreement and the Central America–Mexico Free Trade Agreement). Others contain binding commitments on data flows (of all types of data). Examples include the Comprehensive and Progressive Agreement for Trans-Pacific Partnership (CPTPP); the United States, Mexico, and Canada Agreement; and the EU–UK Trade and Cooperation Agreement. Importantly, CPTPP and the United States, Mexico, and Canada Agreement balance the obligation to allow international data flows with the obligation of countries to institute nondiscriminatory mechanisms to protect the privacy of consumers and to protect consumers from fraud.

The Regional Comprehensive Economic Partnership (RCEP) has provisions on data flows and data localization that go beyond those in previous agreements between RCEP countries but are less stringent than those in the CPTPP.[a] RCEP stipulates free flows of data and no forced data localization but allows for broader exceptions than in the CPTPP for measures to achieve public policy objectives and those for protection of essential security interests, which cannot be challenged by other member countries.

Digital economy framework agreements are treaties that promote cooperation in regulatory approaches in areas such as articficial intelligence and electronic identification, ensuring trusted cross-border data flows in compliance with data protection rules and other public policy objectives and promoting information exchange and cooperation in the field of cybersecurity.

a. RCEP is the world's largest free trade agreement, accounting for about 30 percent of global gross domestic product. For more information, refer to the Singapore Ministry of Trade and Information's web page on the agreement, https://www.mti.gov.sg/Trade/Free-Trade-Agreements/RCEP.

Second, enhance *regulatory predictability in the region through transaction-level measures*:

- Achieve better interoperability across data protection regimes by
  o Increasing the uptake and use of standard contractual clauses such as the ASEAN Model Contractual Clauses to promote legal interoperability across regulatory regimes; and

o Adopting accountability-based mechanisms such as the CBPRs System to support further harmonization of data protection regimes and practices as a complement to national legal obligations and the Model Contractual Clauses.

- Develop capacity and raise awareness to help organizations—particularly micro, small, and medium enterprises—to understand their roles and responsibility in protecting personal data, their regulatory obligations for different jurisdictions, and the purposes and effects of the Model Contractual Clauses and the CBPRs.

- Consider the development of cross-border dataspaces and greater use of data sandboxes, which provide trusted environments for the trusted and secure transfer of data.

Decarbonizing maritime transportation services

International maritime transportation is the backbone of international trade: more than 80 percent of global trade by volume is transported by sea (UNCTAD 2022). In 2021, developing countries accounted for 55 percent of seaborne exports and 61 percent of imports by sea globally. Maritime transportation is particularly important to EAP economies. According to UNCTAD (2022), Asia remains the main loading and discharge center, turning about 42 percent and 64 percent of all global exports and imports, respectively. At the same time, shipping accounts for a significant and growing share of global anthropogenic greenhouse gas (GHG) emissions, contributing about 3 percent of total GHG emissions in 2018 (IMO 2020).

In July 2023, the 176 member states of the International Maritime Organization (IMO) unanimously adopted the 2023 IMO GHG Strategy (IMO 2023). This landmark agreement replaces the 2018 Initial Strategy and significantly strengthens shipping's GHG reduction targets. Concretely, IMO member states agreed to reach net-zero GHG emissions from international shipping by or around 2050, including interim checkpoints of 20–30 percent reductions by 2030 and 70–80 percent by 2040 (figure 4.14).

In addition, IMO's ambition to make zero- or near-zero-GHG energy, fuels, and technologies 5–10 percent of shipping's energy mix by 2030 will significantly change the wider fuel landscape. Annually, international shipping combusts about 300 million tons of mainly heavy and distillate fuel oils, which would need to be gradually replaced (IMO 2020). The most promising candidates for fuel types with zero- or near-zero-GHG emissions are hydrogen-derived fuels produced with renewable energy, such as green ammonia or green methanol (Englert et al. 2021). Meeting the new reduction targets also requires improving the energy efficiency of the fleet, which can be achieved through operational measures, such as speed reductions, or technical measures, such as modifications to the ship's power plant, hull, and propeller.

Meeting IMO's GHG targets will require robust policies. IMO plans to develop a basket of policy measures consisting of both technical and economic elements. The technical element will be a marine fuel standard mandating the phased reduction of the GHG intensity of the energy used onboard ships. This standard is expected to be key to driving the timely and effective uptake of zero- and near-zero-GHG fuels. The specific nature of the economic element still needs to be defined but will likely be based on GHG emissions pricing, such as a carbon price on ship emissions.

FIGURE 4.14   Internatonal shipping aims to reach net-zero GHG emissions by 2050

GHG emissions reductions in international shipping, actual and targets, 2008–50

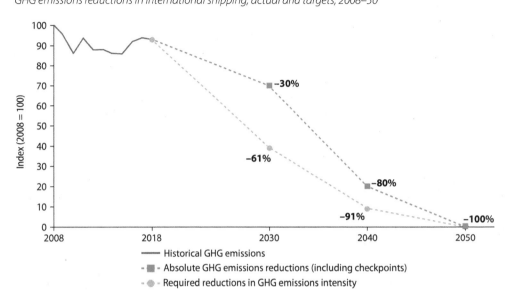

Source: Englert et al. 2023.
Note: GHG = greenhouse gas.

The latter would represent the very first carbon pricing instrument covering all GHG emissions of an entire sector globally. These measures are to be adopted in 2025, with an envisaged entry into force in 2027.

The key role of an economic measure for international shipping is to incentivize reductions in GHG emissions in shipping. According to several studies, economic instruments like carbon pricing have the potential to stimulate the production and uptake of alternative fuels and promote improvements in technical and operational measures on vessels (ITF 2022; Parry et al. 2018). In addition to yielding GHG emissions abatements, these instruments can generate considerable revenues. Estimates indicate that a carbon price applied to international shipping could raise up to US$3.7 trillion by 2050, depending on assumptions related to the carbon price and the emissions reduction trajectory, among others (Baresic et al. 2022; Mærsk Mc-Kinney Møller Center for Zero Carbon Shipping 2021). This amount corresponds to an annual average of US$40 billion–US$60 billion between 2025 and 2050 in collected carbon revenues (Dominioni et al. 2022).

Policy measures to reduce GHG emissions from international shipping can affect the trade of remote countries. A United Nations Conference on Trade and Development study suggests that the IMO measure would have a small global impact on maritime logistics costs when contrasted with the typical fluctuations in freight rates (UNCTAD 2021). The global impact on gross domestic product and trade flows could also be considered small. But developing coastal countries, including small island developing states and least developed countries, would experience a more significant reduction in their gross domestic product and import and export flows than developed coastal countries. Furthermore, some measures the IMO is currently considering, namely a marine fuel standard and a GHG emissions pricing instrument, may have more significant impacts than those of the energy efficiency measure the United Nations Conference on Trade and Development previously examined because the IMO measures must drive the uptake of more expensive zero-carbon bunker fuels.

The IMO has agreed that shipping's decarbonization should be just and equitable. Many IMO member states recognize that carbon pricing cannot only reduce GHG emissions but also address equity concerns. It can do so when revenues are strategically channeled back to those countries that may struggle the most with shipping's decarbonization and/or with climate change—most often developing countries, specifically small island developing states and least developed countries. These revenues could also be used to enhance maritime transportation infrastructure and capacity, and lower countries' maritime transportation costs. Another way would be to offset (some of) the negative impacts by channeling carbon revenues to countries affected disproportionately.

Because of their vulnerability to climate change and their interest in reducing emissions, EAP countries have a stake in the IMO discussions. As traders dependent on shipping, they are also likely to be affected by the increase in the cost of transportation. Therefore, they must engage in cooperative discussions to ensure both meaningful climate action and an equitable distribution of the cost of such action.

Annex 4A. Services for the Pacific

For their stage of development, Pacific island countries tend to rely heavily on their services sectors for employment and growth. At the regional level, services contributed 59 percent to gross domestic product in 2018 while employing nearly 48 percent of the labor force, although employment varied across the region and the definition of services might vary across data sources (figure 4A.1). Many Pacific island countries have become heavily vested in services and even maintain a services trade surplus, allowing services to emerge as drivers of growth in the Pacific. Imports, however, are often just as crucial to growth. In the Pacific, services import composition is mostly concentrated in travel services (49.9 percent of total services imports in 2018), followed by transportation services (20.4 percent), and other business services (13 percent) (figure 4A.2).

Growth of services industries in the Pacific is limited by both the scale of operation and the scale of internal markets, thus causing challenges in ensuring effective competition. In some circumstances, small market sizes can force concentration to dangerously high levels, thereby increasing the risk that providers can create monopolies, form cartels, or abuse their dominance in some other way.

Opening the Pacific to trade in services can help alleviate artificially induced constraints on market size by expanding the scope and intensity of competition or by easing market concentration. However, many countries in the region have various hidden restrictions that limit the ability of firms to operate across borders in an integrated market of any significant scale.

FIGURE 4A.1   Contribution of services to employment varies across Pacific island countries

Employment, by sector, Pacific island countries

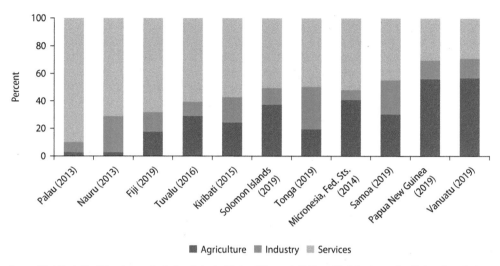

Sources: World Bank, World Development Indicators; Federated States of Micronesia Division of Statistics; International Labour Organization.

FIGURE 4A.2   Pacific island countries' services imports are concentrated in transportation, travel, and other business services

Composition of services imports, by services sector, Pacific island countries

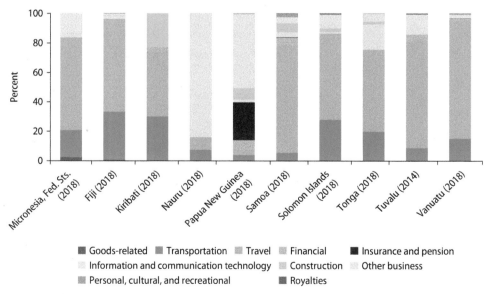

Source: Original figure for this publication using data from the United Nations Conference on Trade and Development.

The Services Trade Restriction Index was recently compiled for the first time for six Pacific island countries, following a regulatory audit of discriminatory measures in the countries' services sectors (figure 4A.3). Although these countries perform moderately on the index scale (relative to the global sample), their much smaller economies mean that the effects of every trade restrictive measure can affect market outcomes to a much greater degree. Consequently, certain restrictions often deepen the countries' isolation and give monopolies to a small set of domestic incumbents. This effect is particularly damaging to the economy in services sectors that either (1) enhance connectivity and integration into the global economy (such as transportation or communication services) or (2) have strong links to other domestic industries (for example, financial and professional services).

For instance, a disaggregated review of the index by transportation subsector and modes of delivery shows that some countries in the region are relatively restricted by one or more measures. Maritime transportation connectivity is especially essential for the region's economy. Yet, countries like the Solomon Islands have effectively maintained a complete sector closure to trade in some of those services. Consequently, the rising cost of port services has been an increasing concern in the Solomon Islands (Kekea 2021), even in the face of state control over pricing. In fact, a comparative analysis of port charges shows that Honiara's became the highest in the region following a port tariff increase in 2015 (table 4A.1). Although state management of port services is an attempt to manage what could otherwise be a private monopoly of an essential facility, there may be more efficient ways to tender out a concession for the operation of wharfage and stevedoring services in a way that

FIGURE 4A.3   Services Trade Restrictiveness Index

STRI outcomes, by broad sector, selected Pacific island countries

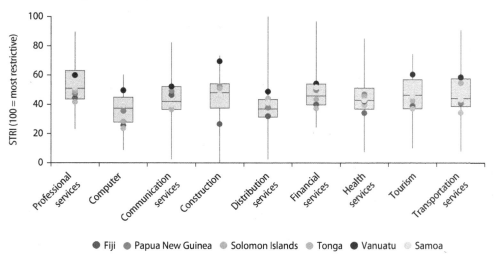

Source: World Bank–World Trade Organization Services Trade Restrictiveness Index (STRI) Database (https://itip-services-worldbank.wto .org/default.aspx).

TABLE 4A.1   Port charges, selected Pacific island countries

Port, country	Amount (US$)
Honiara, Solomon Islands	17,266.19
Lae, Papua New Guinea	10,471.55
Lautoka, Fiji	10,157.28
Nuku'alofa, Tonga	9,277.93
Port Vila, Vanuatu	5,581.99
Apia, Samoa	5,050.21
Average not including Honiara	8,107.79

Source: ADB 2016.

maintains ownership of the facility, while better controlling the cost of their services (ADB 2007).

Air transportation also plays an important role in the connectivity of the region given the geographical conditions of many islands and the links to tourism. However, air transportation services in the region face similar restrictions to foreign entry. For instance, Fiji's Civil Aviation Act of 2012 prevents majority ownership of foreign investors in locally incorporated air passenger companies. Although they do not necessarily prohibit foreign provision of international connections (foreign airlines can still fly into the country), these rules effectively protect domestic routes from foreign competition, leaving such routes to be serviced only by domestic carriers. Because the serviceable market is constrained by regulation, the average fixed costs of establishing an airline to service this market are high. Pro-competitive efforts to liberalize the aviation sector would expand the scope of the market demand by eliminating discriminatory measures imposed on foreign investors. Such efforts were taken seriously in 2003 with the Pacific islands Air Service Agreement, designed to establish a framework for gradual integration of aviation services, but these efforts stalled. Establishing a unified market would increase market size, which could make investment scale less of a binding constraint to the provision of new service.

Retail distribution services represent another sector that is both geared toward domestic markets and subject to many of the scale challenges when the market is limited to national boundaries. For instance, Samoa's Foreign Investment Act of 2000 specifically reserves grocery retailing for domestic investors. However, these foreign investment restrictions in retail distribution can impede competition, which might have led to the food price inflation that far outstrips general inflation and is much higher than in other Pacific island countries.

Last, in the travel sector, attempts to shield small domestic accommodation providers from competition may prevent the emergence of high-value tourism models. For instance, in Vanuatu, investment in accommodation service providers with less than 10 rooms is reserved for domestic investors. Because the prevalent tourism models in the country focus on cruise or sun and sand tourism—models characterized by large-scale investment in accommodation and vertical integration of services—these restrictions have not affected the incumbent industries. However, rules that restrict the scale of investment in accommodation services may inhibit the emergence of higher-value tourism segments. For instance, accommodation services for experiential tourism segments are driven by product uniqueness and ability to service the specific interest of a target niche (scuba, sailing, and so on). Foreign investment in small-scale accommodations (relative to the sun and sand resorts) is often more capable than local investors of servicing these demands. However, the investment restriction on this service sector may create diseconomies of scale, which could harm both the competitive position and product quality for such niche segments. Therefore, the presence of an investment restriction may be preventing the emergence of high-value tourism clusters that could generate improved sectoral productivity.

All these examples reveal how enhanced openness to trade in services could help Pacific island countries improve both market efficiency and competitiveness. The political economy constraints, posed by the incumbents, could be addressed by a sequencing of reforms that starts from the areas where the local businesses' potential losses are likely to be smaller.

Notes

1. By contrast, the following sectors will continue to be categorized as public utilities and thus subjected to the 40 percent foreign ownership cap: distribution and transmission of electricity, petroleum pipeline transmission systems, water distribution systems, seaports, and public utility vehicles.
2. This section is based on World Bank (2023) and World Bank (forthcoming).
3. World Bank and Seoul Center for Finance and Innovation. 2023. "Digital Conglomerates in East Asia: Navigating Competition Policy Challenges." Report. World Bank and Ministry of Economy and Finance, Seoul. World Bank staff analysis based on 2012–21 mergers and acquisitions data from Refinitiv (previously Thompson Reuters).
4. This may not be the case when behavioral advertising is the sector's prevailing business model (Stucke and Ezrachi 2020).
5. ASEAN member countries are Brunei Darussalam, Cambodia, Indonesia, Lao People's Democratic Republic, Malaysia, Myanmar, the Philippines, Singapore, Thailand, and Viet Nam.
6. For a broader discussion about fintech and its regulatory implications, refer to Feyer, Natarajan, and Saal (2023).
7. Refer to Di Gropello (2012) for a more detailed discussion of higher education policy in EAP; refer to Arnhold and Bassett (2021) and Ferreyra et al. (2017) for an examination of higher education policy in other regions.
8. Refer to Fielden and LaRocque (2008) and Di Gropello (2012) for a fuller discussion of governance and regulatory issues in tertiary education in EAP.
9. The proportion of the population owning a smartphone reached about 80 percent in the EAP region in 2021, compared to just over 50 percent in 2016, ranking at the top among the low- and middle- income countries (GSMA 2022).

References

ADB (Asian Development Bank). 2007. "Oceanic Voyages: Shipping in the Pacific." Pacific Studies Series, ADB, Manila.

ADB (Asian Development Bank). 2016. "Consultant Report on Regional Ports Pricing and Productivity Comparative Study." REG: Private Sector Development Initiative Phase III, Sydney.

Afkar, R., T. Béteille, M. E. Breeding, T. Linden, A. D. Mason, A. Mattoo, T. Pfutze, L. M. Sondergaard, and N. Yarrow. 2023. *Fixing the Foundation: Teachers and Basic Education in East Asia and Pacific*. East Asia and Pacific Regional Report. Washington, DC: World Bank.

Almeida, R., J. R. Behrman, and D. A. Robalino. 2012. *The Right Skills for the Job? Rethinking Training Policies for Workers*. Washington, DC: World Bank.

Amanta, F. 2022. "Unpacking Indonesia's Digital Accessibility." Center for Indonesian Policy Studies, November 16, 2022. https://www.cips-indonesia.org/post/opinion-unpacking -indonesia-s-digital-accessibility.

Arnhold, N., and R. M. Bassett. 2021. "Steering Tertiary Education: Toward Resilient Systems that Deliver for All." World Bank, Washington, DC.

Ashton, D., F. Green, J. Sung, and D. James. 2002. "The Evolution of Education and Training Strategies in Singapore, Taiwan and South Korea: A Development Model of Skill Formation." *Journal of Education and Work* 15 (1): 5–30.

Banh, T. H, T. H. Dao, and P. Glewwe. 2024. "An Investigation of the Decline in the Returns to Higher Education in Vietnam." Unpublished working paper.

Baresic, D., I. Rojon, A. Shaw, and N. Rehmatulla. 2022. "Closing the Gap: An Overview of the Policy Options to Close the Competitiveness Gap and Enable an Equitable Zero-Emission Fuel Transition in Shipping." UMAS, London.

Barr, N. 2014. "Income Contingent Loans and Higher Education Financing: Theory and Practice." In *Income Contingent Loans*, edited by B. Chapman, T. Higgins, and J. E. Stiglitz. International Economic Association Series. Palgrave Macmillan, London.

Barzelay, A. 2022. "Landscape Survey of Personal Data Protection Enforcement Decisions." Unpublished working paper.

Chapman, B., and L. Dearden. 2022. "Income-Contingent Loans in Higher Education Financing." IZA World of Labor, Institute of Labor Economics.

Cunningham, W., H. E. Moroz, N. Muller, and A. V. Solatorio. 2021. "The Demand for Digital and Complementary Skills in Southeast Asia." Policy Research Working Paper 10070, World Bank, Washington, DC.

Di Gropello, E. 2012. "Putting Higher Education to Work: Skills and Research for Growth in East Asia." World Bank East Asia and Pacific Regional Report, World Bank, Washington, DC.

Dominioni, G., D. Englert, R. Salgmann, and J. Brown. 2022. "Carbon Revenues from International Shipping: Enabling an Effective and Equitable Energy Transition." Technical Paper, World Bank, Washington DC.

Englert, D., A. Losos, C. Raucci, and T. Smith. 2021. "Volume 1: The Potential of Zero-Carbon Bunker Fuels in Developing Countries." World Bank, Washington, DC.

Englert, D., I. Rojon, R. Salgmann, and S. Sulikova. 2023. "A New Climate Deal for Shipping: Three Decades to Zero." *World Bank Blogs*, July 23, 2023. https://blogs.worldbank.org/en /transport/new-climate-deal-shipping-three-decades-zero.

Ferreyra, M. M., C. Avitabile, J. Botero Álvarez, H. Paz, and F. Urzúa. 2017. *At a Crossroads: Higher Education in Latin America*. Directions in Development, Human Development. Washington, DC: World Bank.

Feyer, E., H. Natarajan, and M. Saal. 2023. *Fintech and the Future of Finance: Market and Policy Implications*. Washington, DC: World Bank.

Fielden, J., and N. LaRocque. 2008. "The Evolving Regulatory Context for Private Education in Emerging Economies." Education Working Paper 14, World Bank, Washington, DC.

Gómez, M. F. V. 2021. "The Relationship between Network Performance and Customer Satisfaction in the Philippines." Ookla, December 7, 2020. https://www.ookla.com/articles /network-performance-customer-satisfaction-philippines-q2-q3-2020.

Grab. 2018. "Grab Merges with Uber in Southeast Asia." Press release, March 26, 2018. https:// www.grab.com/sg/press/business/grab-merges-with-uber-in-southeast-asia/.

Hanushek, E., Y. Wang, and L. Zhang. 2023. "Understanding Trends in Chinese Skill Premiums, 2007–2018." NBER Working Paper 31367, National Bureau of Economic Research, Cambridge, MA.

IMO (International Maritime Organization). 2020. "Fourth IMO Greenhouse Gas Study 2020." IMO, London.

IMO (International Maritime Organization). 2023. "2023 IMO Strategy on Reduction of GHG Emissions from Ships." Resolution MEPC.377(80), Marine Environment Protection Committee, IMO, London.

ITF (International Transport Forum). 2022. "Carbon Pricing in Shipping." International Transport Forum Policy Paper 110, OECD Publishing, Paris.

Kekea, G. 2021. "Ports Authority to Increase Entrance Fees for Domestic Port." *Solomon Times*, September 22, 2021. https://www.solomontimes.com/news/ports-authority-to -increase-entrance-fees-for-domestic-port/11116.

Mærsk Mc-Kinney Møller Center for Zero Carbon Shipping. 2021. "Options Paper on Market-Based Measures." Mærsk Mc-Kinney Møller Center for Zero Carbon Shipping, Copenhagen. https://cms.zerocarbonshipping.com/media/uploads/documents/MBM-Options-Paper-Final .pdf.

Montenegro, C. E., and H. A. Patrinos. 2023. "A Data Set of Comparable Estimates of the Private Rate of Return to Schooling in the World, 1970–2014." *International Journal of Manpower* 44 (6): 1248–68.

Motta M., and M. Peitz. 2021. "Big Tech Mergers." *Information Economics and Policy* 54 (March): 100868.

Nyman, S., A. B. Ruival, and T. Begazo. 2023. "Ex Ante Competition Regulation of Digital Platforms." Unpublished working paper.

Parry, I., D. Heine, K. Kizzier, and T. Smith. 2018. "Carbon Taxation for International Maritime Fuels: Assessing the Options." IMF Working Paper WP/18/203, International Monetary Fund, Washington, DC.

Stucke, M. E., and A. Ezrachi. 2020. *Competition Overdose: How Free Market Mythology Transformed Us from Citizen Kings to Market Servants*. First edition. New York: Harper Business.

UNCTAD (United Nations Conference on Trade and Development). 2021. *UNCTAD Assessment of the Impact of the IMO Short-Term GHG Reduction Measure on States*. Geneva: United Nations.

UNCTAD (United Nations Conference on Trade and Development). 2022. *Review of Maritime Transport 2022*. New York and Geneva: UNCTAD.

World Bank. 2011. *Putting Higher Education to Work: Skills and Research for Growth in East Asia*. Washington, DC: World Bank.

World Bank. 2021a. "Beyond Unicorns: Harnessing Digital Technologies for Inclusion in Indonesia." World Bank, Washington, DC.

World Bank. 2021b. *World Development Report 2020: Data for Better Lives*. Washington, DC: World Bank.

World Bank. 2023. "Digital Platforms in Developing Economies: Options to Mitigate Competition Risks." World Bank, Washington, DC.

World Bank. 2024. "Measuring and Addressing Inequality in Malaysia." Unpublished working paper.

World Bank. Forthcoming. "Navigating Conglomerates in East Asia's Digital Markets: Enhancing Competition Policies and Regulations." World Bank, Washington, DC.

Looking Forward: A Virtuous Cycle of Opportunity and Capacity   5

Conclusion

This report has argued that technological change, particularly digitalization, and policy reforms can unleash a virtuous cycle of opportunity and capacity. Drawing on firm-level data from the Philippines and Viet Nam, it first showed that technology and reforms stimulate productivity improvements, both in services sectors and in the manufacturing sectors using those services. Second, using microlevel data from Indonesia, the analysis found that jobs in digitalized services sectors pay higher wages and offer opportunities for the more skilled—showing, in parallel, that technology and reforms can improve access to education, health, and finance, thus equipping people with the capacity to seize the new opportunities. Finally, the report discussed how unleashing the virtuous dynamic requires striking three type of policy balances: (1) between liberalization and regulation, (2) between the state and the market in creating the infrastructure and skills needed to take advantage of opportunities, and (3) between unilateral domestic reform and cooperative international action to address services market failures that have a transborder dimension.

Looking ahead, it will be important for countries in the East Asia and Pacific region to pursue development strategies that are attuned to the interplay between opportunity and capacity. The simple conceptual framework reported in figure 5.1 illustrates the point. This figure makes more explicit the alternative dynamic paths flowing from the interplay depicted in figure 1.9 (the organizing framework set out in chapter 1). Starting from positions of relative underdevelopment (the "LL" quadrant in figure 5.1), countries might pursue development strategies that result in a significant increase in opportunities, without at the same time creating the commensurate capacities (HL quadrant). This path would potentially lead to shortages of skills. Alternatively, countries could concentrate too much on the creation of capacities, without at the same time fostering the emergence of the opportunities needed to suitably employ those new capacities (LH quadrant). This path would potentially lead to underuse (reflected, for instance, in high youth unemployment rates). Clearly, the optimal situation is one in which the policies put in place facilitate a balanced

FIGURE 5.1 **The need for a balanced development of opportunities and capacities**

High capacities

| Potential underuse **LH** | Virtuous development **HH** |

Low opportunities ←→ High opportunities

| **LL** Underdevelopment | **HL** Potential shortages |

Low capacities

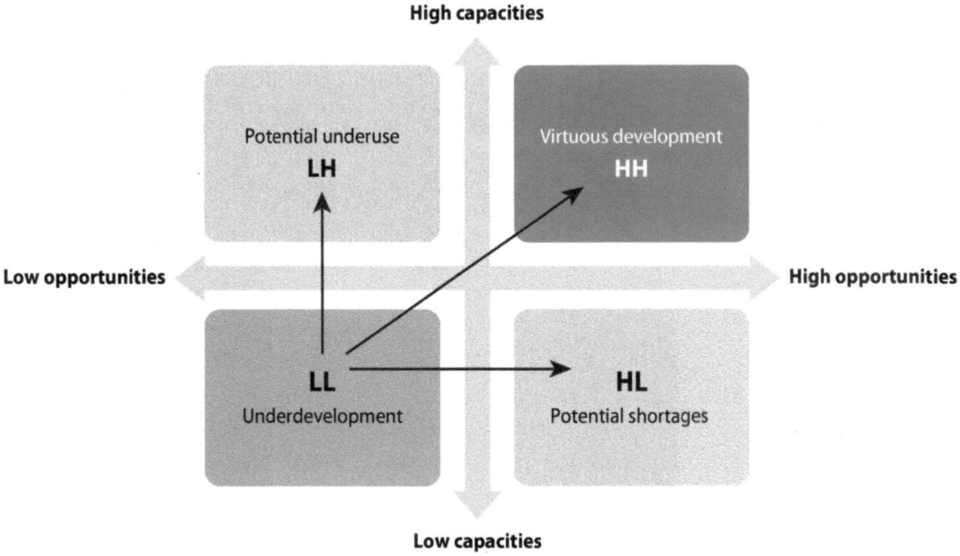

Source: Original figure for this publication using data from the World Bank.

development of opportunities and capacities—that is, from quadrant LL, they travel to quadrant HH.

Considering specific measures of capacities and opportunities lends some empirical content to the conceptual framework in figure 5.1. Focusing on skills, tertiary enrollment rates can be considered a measure of capacity, and the skilled labor content of export (which has the virtue of being driven by external demand) as a measure of opportunities. Figure 5.2 presents a scatter plot along these two dimensions for all economies with available data on these measures. The horizontal axis reports the skilled labor content of exports in 2014 (the last year for which information is available from Calì et al. 2016) and the vertical axis the average tertiary enrollment rate in 2012–16 (to maximize economy coverage). The global sample medians of these measures are also plotted as black dashed lines.

As figure 5.2 highlights, in 2014, countries such as Australia, Japan, and Singapore had both opportunities and capacities above the global median. Countries like Mongolia and Thailand exhibited tertiary enrollment rates above the sample median but demand for skilled labor below the sample median. By contrast, countries like Ghana, India, and Morocco displayed demand for skills above the sample median but enrollment rates below the median. Finally, countries such as Cambodia and Indonesia displayed both capacities and opportunities below the median.

Although figure 5.2 presents a static picture from the past as a purely expositional device, looking more closely at the experiences of China, the Republic of Korea, and Viet Nam illustrates how dynamic differences in the pace of expansion of the supply of and demand for skills can lead to imbalances. These examples highlight the important role of policies that expand both supply and demand.

FIGURE 5.2   **An illustration of opportunities and capacities in selected economies**

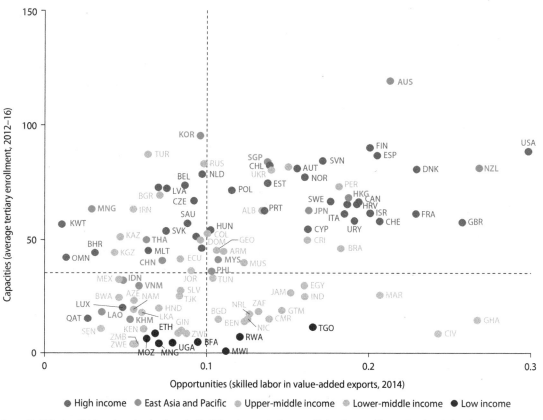

Source: World Bank staff elaboration using data from the World Development Indicators and Labor Content of Exports Database (Calì et al. 2016).

After achieving universal basic education, Korea expanded higher education to create the foundations of high-technology sectors. Korea's tertiary education enrollment is now among the highest in the world, with a high share of tertiary graduates with a strong focus on science, technology, engineering, and mathematics (STEM). The government has systematically embedded human capital development in national economic planning and forges public-private partnerships to ensure coordination of the supply and demand of skilled labor. The expansion of tertiary education has been instrumental in enabling innovation- and technology-led manufacturing and modern services. The incentives to invest in tertiary education have remained strong: the returns rose in the 1980s, fell somewhat in the 1990s, and rose again through the early 2000s.

Despite those positive results, some concerns have recently arisen in Korea about overinvestment in tertiary education and skill mismatches, because tertiary graduates are taking on jobs that may not require a four-year (or more) college degree. The government has implemented measures to tie technical education and shorter-cycle tertiary programs to the economy's needs in skilled trade jobs. These measures include the creation of the Meister Schools, which offer apprenticeship workplace-based training in vocational high schools and community and junior colleges. These programs have become increasingly attractive because they have strong links to industry and can facilitate the transition to university for those who choose to pursue a college degree. Korea offers positive lessons on how the government can systematically steer skills development to further long-term economic development through a balanced approach to skills investments and public-private partnerships that avoid coordination failures.

Prioritizing universal basic education, China embarked on one of the fastest and largest expansions of tertiary education in world history (World Bank Group and the Development Research Center of the State Council, the People's Republic of China 2019). Tertiary education enrollment—with a strong focus on science, technology, engineering, and mathematics education—has more than tripled in the country since 2000. The tertiary gross enrollment rate now surpasses 60 percent, and more than 40 percent of college graduates pursue STEM fields.

The boost in the supply of tertiary graduates was met with a gradually increasing demand for skills. China's rise to become a global manufacturing powerhouse during the 1990s and 2000s expanded the demand for its abundant stock of labor with low and middle skills, pulling many workers out of agriculture.[1] Labor costs then shifted China's global comparative advantage, and the induced automation eroded the role of manufacturing as a source of intermediate-skill jobs. China responded by steering domestic and foreign investment to new technologies. For instance, China's share of all new industrial robots installed worldwide was 0.4 percent in 2000, rose to 12.0 percent in 2010, and surpassed 50.0 percent in 2022 to make China the world's largest market for newly deployed industrial robots (Müller 2023). The expansion of tertiary graduates gradually enabled endogenous technological upgrading and productivity growth in manufacturing, which in turn boosted the demand for skilled labor (Che and Zhang 2018; Feng and Xia 2022). The returns to tertiary education almost tripled from the early 1990s to the late 2000s, as the supply expansion lagged the increased demand (Ma and Iwasaki 2021).

Starting in the mid-2010s, and especially in more recent years, signs indicate that the supply of tertiary graduates began to outpace demand. The share of employed workers with some college education rose from 5.6 percent in 2001 to 19.1 percent in 2018 but increased rapidly only after 2009 when a large inflow of college graduates entered the labor market (Hanushek, Wang, and Zhang 2023). A recent study estimates that the average return to higher education in China decreased sharply from 68 percent in 2007 to 41 percent in 2013, recovering only slightly to 49 percent by 2018. During this period, China saw a restructuring of the economy with a steady rise in the share of employment in services. The expanding services sectors absorbed a large share of the increased supply of college graduates. Those graduates who secured jobs in skills-intensive services (such as finance, e-commerce, professional services, consulting, and real estate) fared better than those employed in less skills-intensive sectors (sales, wholesale, and retail services). In fact, significant variation exists in the returns to college: many graduates face nearly zero or even negative returns (World Bank Group and the Development Research Center of the State Council, the People's Republic of China 2019). Starting in 2020, unemployment among college graduates surged because of COVID-19 lockdowns and regulatory interventions that hit services harder, and because weak labor market conditions led to increased enrollment and a subsequent spike in the inflow of graduates in 2022.

China's experience also illustrates how investments and policies to foster the supply of skills can enable a transition to higher-value-added manufacturing and services. At the same time, it reflects the need to carefully manage surges in the supply of college-educated workers and maintain an environment that preserves a strong demand for skilled workers.

Finally, Viet Nam has also made significant investments in tertiary education, building on strong foundations of nearly universal access to quality basic education. Expansion has been somewhat slower than in China and Korea, but notably faster over the last decade. The tertiary education enrollment rate rose from 3 percent in 1990 to 24 percent in 2010 and to 42 percent in 2022. The share of workers with a college education increased from about 3 percent in 1993 to 12 percent in 2010 and to 17 percent in 2020. A recent study estimates that the average return to a university education rose moderately in the 2000s and then fell from 45 percent in 2013 to 32 percent in 2019.

Given the small increase in the share of college-educated workers during that period, the faltering returns to education in Viet Nam are possibly associated with a relatively weak demand for skills. Viet Nam's integration into global value chains, after joining the World Trade Organization in 2007, has been largely into exports with a relatively low skill content. This expansion in low-skill jobs has been accentuated by the country's absorbing a substantial share of manufacturing displaced from China in recent years. As new college graduates enter the labor market, the downward pressure on the college wage premium is bound to increase, thus eroding the incentives to invest in higher education. The experiences of China and Korea show the need to offset this trend by harnessing the potential of new technologies in manufacturing and skilled-intensive services to foster the demand for skills (refer to "out of the box" 3.1 in chapter 3).

These three examples illustrate the importance of pursuing balanced policies in which the enhancement of endowments shapes comparative advantage and the evolution of comparative advantage incentivizes the enhancement of endowments, generating a virtuous cycle that powers development.

Note

1. In contrast, high-income economies and most middle-income countries saw a sharp fall in the share of employment with middle skills (World Bank 2016).

References

Calì, M., J. Francois, C. Hollweg, M. Manchin, D. A. Oberdabernig, H. Rojas-Romagosa, S. Rubinova, and P. Tomberger. 2016. "The Labor Content of Exports Database." Policy Research Working Paper 7615, World Bank, Washington, DC.

Che, Y., and L. Zhang. 2018. "Human Capital, Technology Adoption and Firm Performance: Impacts of China's Higher Education Expansion in the Late 1990s." *Economic Journal* 128 (614): 2282–320.

Feng, S., and X. Xia. 2022. "Heterogeneous Firm Responses to Increases in High-Skilled Workers: Evidence from China's College Enrollment Expansion." *China Economic Review* 73 (June): 101791.

Hanushek, E., Y. Wang, and L. Zhang. 2023. "Understanding Trends in Chinese Skill Premiums, 2007–2018." Working Paper 31367, National Bureau of Economic Research, Cambridge, MA.

Ma, X., and I. Iwasaki. 2021. "Return to Schooling in China: A Large Meta-analysis." *Education Economics* 29 (4): 379–410.

Müller, Christopher. 2023. *World Robotics 2023—Industrial Robots*. Frankfurt am Main, Germany: VDMA Services GmbH.

World Bank. 2016. *World Development Report 2016: Digital Dividends*. Washington, DC: World Bank.

World Bank Group and the Development Research Center of the State Council, the People's Republic of China. 2019. *Innovative China: New Drivers of Growth*. Washington, DC: World Bank.

Appendix A. Share of Manufacturing and Services in Value Added and Employment

FIGURE A.1   Share of manufacturing and services in value added, selected EAP countries, 2010 vs. 2021

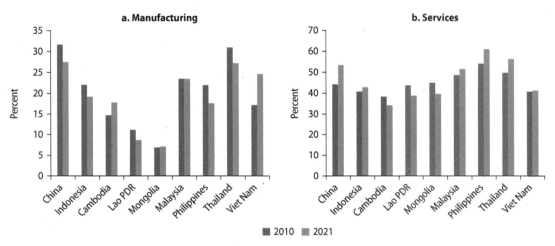

Source: International Labour Organization Department of Statistics, ILOSTAT; World Bank, World Development Indicators.
Note: EAP = East Asia and Pacific.

FIGURE A.2   Share of manufacturing and services in employment, selected EAP countries, 2010 vs. 2021

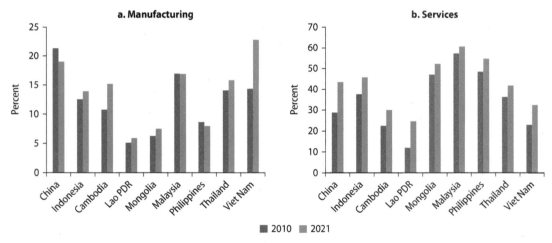

Source: International Labour Organization Department of Statistics, ILOSTAT; World Bank, World Development Indicators.
Note: EAP = East Asia and Pacific.

Appendix B. Growth of Foreign Direct Investment Inflows, Cumulative Share by Sector

FIGURE B.1   **Growth of FDI inflows, cumulative share by sector, selected EAP countries**

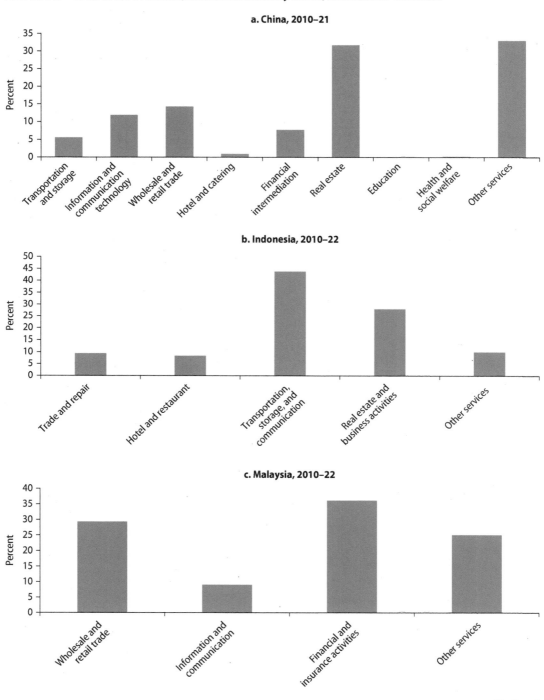

(Continued)

FIGURE B.1   Growth of FDI inflows, cumulative share by sector, selected EAP countries *(Continued)*

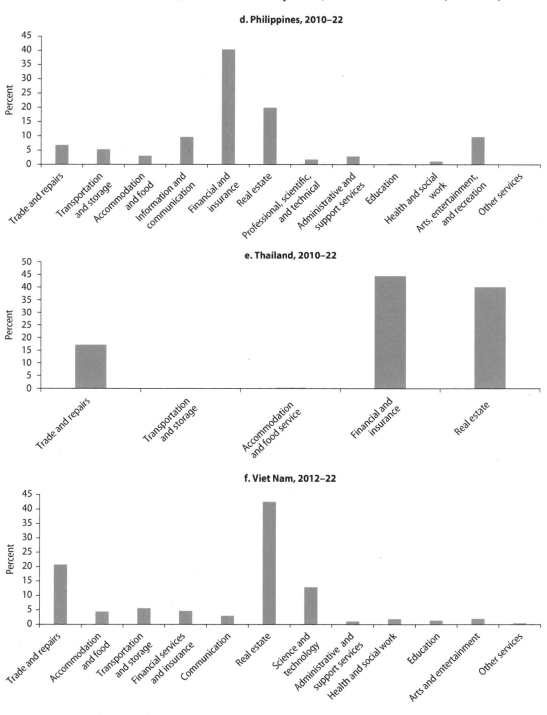

Source: Haver Analytics; World Bank, World Development Indicators.
Note: EAP = East Asia and Pacific; FDI = foreign direct investment.

Appendix C. Growth in Manufacturing Trade, Services Trade, and Digitally Deliverable Services Trade, 2005–21

FIGURE C.1   Growth in trade, by type, selected EAP countries

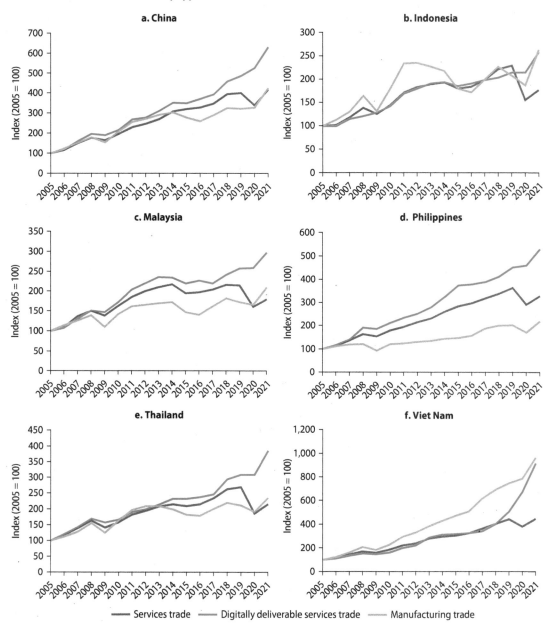

Source: Organisation for Economic Co-operation and Development–World Trade Organization, Balanced Trade in Services (BaTIS) data set.
Note: EAP = East Asia and Pacific.

Appendix D. Definitions of the 44 Digital Subsectors Reported in the FCI Digital Business Database

Subsector	Definition
3D printing	Developing and using 3D printing, or additive manufacturing, refers to the manufacturing process and the technology related to printing a three-dimensional object. This sector encompasses the actual printer as well as software related to 3D printing.
Aerospace tech	Developing and using technology to provide services, research, and innovation related to space flight, aviation, satellite, and space exploration. This subsector includes but is not limited to satellite operations management software, products enabled by satellite connection (such as real-time aerial mapping), spacecraft and aircraft development software, and spatial communication technology.
Agtech	Developing and using digital technologies to enable the agriculture technology value chain, including but not limited to digital agriculture software and hardware (sensors, imagery, precision ag); mixed and integrated agricultural innovation; plat and crop science; animal and livestock science; post-farm agriculture value chain (agri marketplace, delivery, logistics, supply chain innovation); and agricultural waste management.
Artificial intelligence	Developing and using technology for machines to autonomously learn and act through data analytics. This sector will inevitably be closely related to big data and analytics, since AI utilizes a large quantity of data to perform their given functions.
Big data and analytics	Developing and using technology for recording, collection, distribution, and use of large volumes of data. Big data refers to data that is too large, too fast, or too complex to process using traditional methods. This sector includes firms that use data as a service, data analysis and visualization services, and data collection services.
Biotech	Developing and using biotechnology to create products that are dependent upon developing and creating new products by using and manipulating biological systems and living organisms. This subsector includes firms developing databases for biotech research and IoT devices for biotech.

(Continued)

Subsector	Definition
Blockchain and cryptocurrency	Developing and using technology to use blockchain applications and distribute ledger technology. This subsector includes but is not limited to firms using smart contracts, crowd funding, supply chain auditing, cryptocurrency, identity management, intellectual property, file storage, etc. The cryptocurrency space includes companies providing services or developing technology related to the exchange, storage, facilitation of payments, and securing cryptocurrency.
Business management tech	Developing and using technology to improve business operations. This subsector includes but is not limited to operations management/optimization software, customer relations management, customer service tools, enterprise resource planning products, and corporate digitization consulting.
Civic tech	Developing and using technology to improve and aid the relationship between civil society, governmental functions, and humanitarian well-being. This subsector includes but is not limited to government management systems, data analytics on political and governance processes, taxation management, civil society reporting systems, and monitoring products and services.
Clean tech	Developing and using technology to improve the creation, distribution, use, and monitoring of clean and sustainable products and services. This subsector includes but is not limited to digitally-enabled clean energy products and services, sustainable product e-commerce, clean technology logistics technology, and recycling and waste management technology.
Construction tech	Developing and using technology to improve construction value chain. This subsector includes but is not limited to construction operation management software, construction safety IoT services, and construction logistics software.
Digital media	Developing and using technology to improve the creation, editing, storage, access, distribution, publishing, analysis, and delivery of media in digital settings. This subsector includes but is not limited to digital journalism, social media, e-media searching and subscription platforms, and publishing logistics management products and services.
Drones	Developing the technology, and using, servicing, and delivering automated or remote-controlled mechanical devices and technology, including unmanned aerial vehicles, subsea vehicles, and land vehicles.
E-commerce	Developing and using digital technology to facilitate and improve the sale of products over internet networks. The Bureau of Economic Analysis considers e-commerce to include digitally ordered, digitally delivered, or platform-enabled transactions. This subsector includes but is not limited to online marketplaces, aggregator e-commerce, e-commerce analytics, e-commerce transactions, and e-commerce logistics.
Edtech	Developing and using technology to enhance teaching, learning, and training both inside and outside of classrooms. This subsector includes but is not limited to learning devices (tablets and interactive "smart" boards), educational institution management systems, virtual learning products and services, remote learning products and services, and instructor and student assistance programs.

(Continued)

Subsector	Definition
Entertainment tech	Developing and using technology to improve the creation, distribution, delivery, analysis, and use of entertainment products and services. This subsector includes but is not limited to e-sports, e-casino, movies, animation studios and gaming (hardware and software) products, music and video streaming platforms and services, arts, music algorithm software, online management of entertainment events, and entertainment-oriented social media.
Fintech	Developing and using technology for financial services that is usually offered by traditional banks. This includes loans, payments, and wealth and investment management, as well as software providers of automating financial processes or products that address the core business needs of financial firms.
Food tech	Developing and using technology to improve food and beverage production, distribution, purchasing, and consumption. This subsector includes but is not limited to restaurant aggregator/ review platforms, food e-marketplaces, and food lifestyle media, as well as prepackaged food subscription firms.
Gig economy	Developing and using technology to connect gig-economy workers to gig-economy opportunities, including the sharing of economic opportunities. This subsector includes but is not limited to freelancer/gig-worker hiring platforms, gig worker workflow management software, and gig worker insurance platforms.
Health tech	Developing and using technology to improve the creation, facilitation, delivery, safety, reliability, and analysis of health care services. This subsector includes but is not limited to telehealth, e-health platforms, pharmatech, technical medical device development, medical laboratory management, and diagnostic algorithm development.
HR tech	Developing and using technology to improve the management, research, analysis, and organization of human resource functions. This subsector includes but is not limited to human resource management software/platform, recruitment algorithms, job posting platforms, employee performance and time tracking, employee training (remote and/or virtual), and reporting tools.
Insurance tech	Developing and using technology to improve the creation, distribution, delivery, use, and analysis of insurance products and services.
Internet of things	Developing, producing, and using internet of things devices (physical objects that are embedded with sensors that monitor, store, and send data for use in the physical space).
Legal tech	Developing and using technology to improve creating, distributing, using, interpreting, organizing, and assessing legal products and services. This subsector includes but is not limited to telelegal services, legal service aggregators, algorithmic legal services, and caseload management solutions.
Logistics tech	Developing and using technology to improve the movement of goods. This subsector includes but is not limited to digital supply chain management, cargo management software, supply chain tracking, and operation management software.

(Continued)

Subsector	Definition
Manufacturing tech	Developing and using technology to improve the operation and management of the manufacturing value chain. This subsector includes but is not limited to automation solutions, "smart" factory products, and data-based production analytics tools.
Marketing tech	Developing and using technology to improve the marketing value chain. This subsector includes but is not limited to digital marketing content creation, digital marketing consultancy, marketing data and analytics, search engine optimization technology, and customer tracking and interaction products and services.
Mining tech	Developing and using technology to improve the mining value chain. This subsector includes but is not limited to seismic data analytics, mining operation optimization, supply chain management software, and risk detection technologies.
Mobility tech	Developing and using technology to improve the movement of people. This subsector includes but is not limited to passenger transportation logistics (for travel by air, train, and automobile), traffic monitoring and tracking, on-demand ride share and hauling (for both motorized and nonmotorized transportation), passenger transportation repair platforms, and online maps.
Nanotech	Developing and using nanotechnology to create products that are dependent upon the ability to manipulate materials at an atomic level, usually due to the materials exhibiting novel properties at the subatomic level.
Pet tech	Developing and using technology to improve products and services for animal and pet care. This subsector includes but is not limited to animal care matching platforms, televet care, animal product e-commerce, animal monitoring IoT and wearables, and animal care social media.
Property tech	Developing and using technology to improve the real estate and property development value chain. This subsector includes but is not limited to property sale and renting platforms, property management software, renter verification software, and smart home applications.
Quantum tech	Developing and using digital technology through quantum computing principals (using Qubits instead of normal computer bits of either 0 or 1). This subsector includes the hardware and software components of quantum computing.
Reality tech	Developing and using technology that provides user experiences in a different reality environment. This includes both virtual and augmented reality.
Robotics	Developing and using technology for remote-controlled mechanical devices, including machines programmed to perform repetitive and precision tasks.
Security tech	Developing and using technology to improve safety and security products and services. This subsector includes but is not limited to cybersecurity-related products and services, security monitoring and security IoTs, and wearables.

(Continued)

Subsector	Definition
Social network	Developing and using technology to enable users to connect and communicate with each other by posting information, comments, messages, and/or images through a dedicated website or application. This subsector includes social media, messaging platforms, services conducted through social media, and content-sharing platforms.
Software and SaaS	Developing and using technology to offer Software as a Service or product. This subsector includes but is not limited to digital infrastructure software, application and web design/coding, industry-specific software, etc.
Tech hardware	Producing or contributing to the process of producing the physical parts of computer, machinery, and related devices that enable digital infrastructure and digital use. Includes firms making or servicing internal and external hardware for devices that enable digital connectivity and software installment.
Telecom	Developing and deploying telecommunication technology to enable digital infrastructure and digital connectivity. This subsector includes but is not limited to telecommunication service providers, telecom infrastructure developers (tech hardware related to broadband and fiber optics), and internet connectivity services (internet and mobile network services) for both individual consumers and businesses.
Travel tech	Developing and using technology to improve the travel and tourism value chain. This subsector includes but is not limited to travel booking platforms, travel review and discovery platforms, and travel security software.
Utilities tech	Developing and using technology to improve the utility value chain, including water and waste management utilities. This subsector includes but is not limited to utility management software, utilities monitoring and tracking services, mobile payment for utilities, leak detection IoTs, technology-enabled toilets, sanitation IoTs, sanitation monitoring tools, and sanitation-related telehealth products and services.
Wearables	Developing and using wearable devices with sensors that collect and analyze data based on the user's activities. This subsector includes firms developing software and hardware related to wearable technology.
Web services	Developing and using technologies to connect users to web-based application and data sources via standard web protocol. This subsector includes but is not limited to hosting services, cloud services, web and application development, web application engineering, and information and communication technology connectivity solution providers.

Source: Zhu et al. 2022.
Note: IoT = internet of things.

Reference

Zhu, T. J., P. Grinsted, H. Song, and M. Velamuri. 2022. *A Spiky Digital Business Landscape: What Can Developing Countries Do?* Washington, DC: World Bank.

Appendix E. Distribution of Venture Capital Funding

FIGURE E.1 Distribution of VC funding, by sector, selected EAP countries, 2022

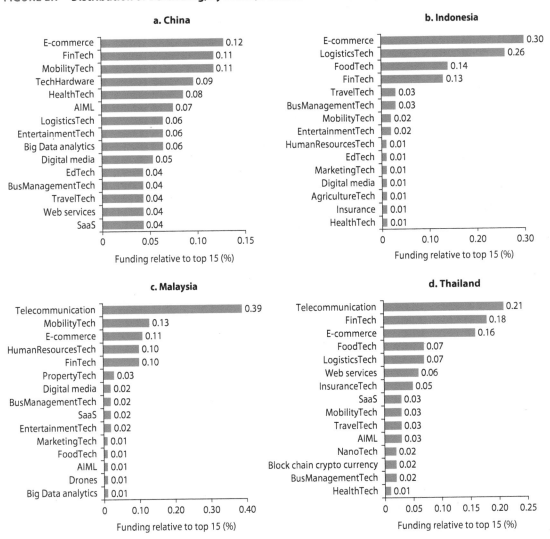

Source: FCI Digital Business Database.
Note: AIML = artificial intelligence and machine learning; FinTech = financial technology; SaaS = Software as a Service; VC = venture capital.

Appendix F. Shaping Tourism Development: The Impact of Digitalization and New Technology Dynamics in East Asia and Pacific

In 2019, travel and tourism generated 9.8 percent of gross domestic product and 184.3 million jobs in East Asia and Pacific (EAP), the region with the highest tourism contribution in absolute terms and the second in relative terms after the Caribbean (WTTC 2023). For some countries in the region, the tourism sector had even higher relevance: Vanuatu, Fiji, and Cambodia derived 32.7, 31.3, and 25.8 percent, respectively, of their gross domestic product from tourism (figure F.1). China, conversely, represented the world's largest outbound tourism market, in terms of both departures and tourism spending, generating a total of US$255 billion for the global economy (UNWTO 2023).

FIGURE F.1 **Tourism as a share of GDP and employment, selected low- and middle-income EAP countries, 2019**

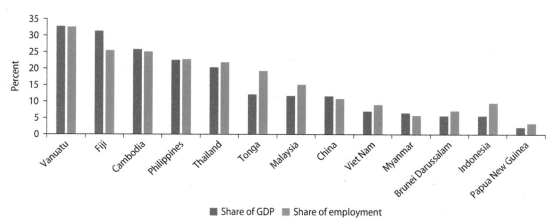

Source: WTTC 2023.
Note: EAP = East Asia and Pacific.

How digital technologies are shaping EAP's tourism industry

The digitalization of tourism distribution channels, driven by online travel agencies and online aggregators, has led to lower consumer prices for airline tickets and hotel rooms. It has simplified product and service comparisons, intensifying market competition and naturally reducing prices (Orlov 2011). In the European Union, online travel agencies lowered average daily hotel rates by €9.40 per room night (7.1 percent reduction) between 2014 and 2019, and contributed 134.1 million additional hotel nights in 2019 (Tourism Economics 2021). Peer-to-peer accommodation platforms that allow individual short-term rentals, like Airbnb and VRBO, have enabled increased choice and competition, and lowered hotel prices. In the 50 largest US cities, hotels experienced a 1.6 percent revenue decrease and up to a 2.8 percent reduction in variable profits because of competition from Airbnb in 2014. The impact varied by segment, location, and date, with a stronger negative effect in cities that have limited hotel capacity (Farronato and Fradkin 2022). Similarly, Airbnb's impact is more noticeable in low-end and Asian markets than in Europe, where hotels can offer more customized and authentic experiences (Yang et al. 2021).

Technology also affects the quality of service in the tourism industry in three ways: review sites, artificial intelligence technology supporting customer service, and greater inclusion for visitors. Reviews have a high impact on consumer decisions. A 2014 survey by TrustYou found that consumers are 3.9 times more likely to book a hotel with higher review scores with equal prices (Ady and Quadri-Felitti 2014). The industry makes use of artificial intelligence technology to support several customer service functions and improve customer experience, including through facial recognition in airports; chatbot software for queries, immediate support, and personalized recommendations; and optimization for recommendations on timing of purchases based on past dynamic pricing trends (Samala et al. 2022). Finally, digital technology can boost inclusiveness and accessibility for visitors with visual, auditory, and cognitive impairments (Michopoulou and Buhalis 2013).

Adoption of digital technologies in EAP's tourism industry

Digital technologies in the tourism sector in EAP are primarily employed for basic front-end business functions, such as advertising. Few firms possess the capacity to conduct end-to-end digital transactions, including property management systems, payments, or digital tools for essential back-end functions such as customer relations management, inventory, finance, and accounting (figure F.2). An (unpublished) business survey conducted by the World Bank across seven Pacific island countries (Fiji, Kiribati, Papua New Guinea, Samoa, the Solomon Islands, Tonga, and Vanuatu) in 2020 revealed that, although most firms had internet access and used social media for business purposes, very few engaged in online sales—and even fewer could execute complete digital transactions, including payments. Data from a World Bank Firm Technology Adoption Survey in Cambodia of 66 accommodation firms showed low levels of technological sophistication in all specific business functions, with a preponderance of either manual systems or the use of standard software, as opposed to specific digital solutions.

FIGURE F.2   Share of hotels using property management systems, Pacific islands and the rest of the world

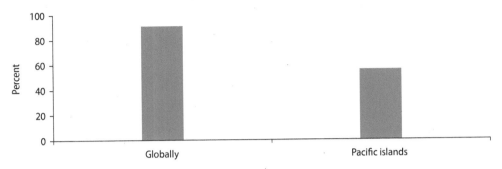

Source: Kovena 2022.

FIGURE F.3   Hotel bookings and occupancy rates, by degree of property management digitalization, with and without property management systems

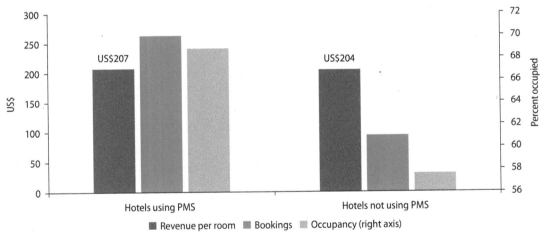

Source: Kovena 2022.
Note: PMS = property management system.

A 2022 survey of more than 130 hotels in EAP found that they paid 211 percent more for international transactions than their peers in other regions and that 54 percent did not offer card payment at the time of booking (Kovena 2022). The same study emphasized the impact that the lack of property management systems has on hotel bookings and occupancy rates (figure F.3).

Low technology adoption reflects multiple factors. Internet penetration in EAP is lower (48 percent) than in the rest of the world (62 percent) (Oxford Economics and PATA 2018). Infrastructure challenges, geographical diversity, or economic barriers could be contributing to that lower penetration. Within the region, there is a clear correlation between internet penetration and average income (Oxford Economics and PATA 2018). Nearly 50 percent of small and medium enterprises and about 30 percent of large firms in emerging EAP economies struggle to digitalize because of financing challenges (Dabla-Norris et al. 2023). Other barriers are a skilled workforce shortage and legal limitations in data protection and intellectual property rights (Sayeh, Dabla-Norris, and Kinda 2023).

Risks associated with the use of digital technologies in tourism

The dominance of digital platforms and solutions in tourism—like Amadeus, Expedia, Priceline Group, Sabre, and Travelport—can bring benefits but also raises concerns due to network effects. These companies hold up to 95–99 percent of market share, potentially generating market inefficiencies that increase costs for businesses and prices for consumers (World Bank n.d.).

The relationship between digital platforms and accommodation providers in the tourism sector raises antitrust and vertical restraint concerns, prevalent because of the sector's reliance on small businesses. Most favored nation clauses used by these platforms can drive up prices for sellers and platform fees, discouraging new entrants and stifling innovation. This issue is reflected in a significant portion of national antitrust cases, accounting for about 13 percent of total cases in developed countries and approximately 10 percent of total cases in developing economies from 2006 to 2019. These cases primarily focus on examining pricing dynamics (World Bank 2021).

References

Ady, M., and D. Quadri-Felitti. 2014. "The Effect of Reviews on Hotel Conversion Rates and Pricing." TrustYou, Munich. https://resources.trustyou.com/c/wp-effect-of-reviews?x=ceEaG6&cn=wp-effect-of-reviews&ct=White%20Paper.

Dabla-Norris, E., T. Kinda, K. Chahande, H. Chain, Y. Chen, A. De Stefani, Y. Kido, F. Qi, and A. Sollaci. 2023. *Accelerating Innovation and Digitalization in Asia to Boost Productivity*. Departmental Paper No 2023/001. Washington, DC: International Monetary Fund.

Farronato, C., and A. Fradkin. 2022. "The Welfare Effects of Peer Entry: The Case of Airbnb and the Accommodation Industry." *American Economic Review* 112 (6): 1782–817.

Kovena. 2022. "Pacific Islands Hotel Insights 2022." Kovena. https://kovena.com/blog/pacific-islands-hotel-insights-2022/.

Michopoulou, E., and D. Buhalis. 2013. "Information Provision for Challenging Markets: The Case of the Accessibility Requiring Market in the Context of Tourism." *Information & Management* 50 (5): 229–39.

Orlov, E. 2011. "How Does the Internet Influence Price Dispersion? Evidence from the Airline Industry." *Journal of Industrial Economics* 59 (1): 21–37.

Oxford Economics and PATA (Pacific Asia Travel Association). 2018. *Data & Digital Platforms: Driving Tourism Growth in Asia Pacific*. Bangkok: Oxford Economics and PATA.

Samala, N., B. S. Katkam, R. S. Bellamkonda, and R. Rodriguez. 2022. "Impact of AI and Robotics in the Tourism Sector: A Critical Insight." *Journal of Tourism Futures* 8 (1): 73–87.

Sayeh, A. M., E. Dabla-Norris, and T. Kinda, T. 2023. "Asia's Productivity Needs a Boost That Digitalization Can Provide." *IMF Blog*, January 9, 2023. https://www.imf.org /en/Blogs/Articles/2023/01/09/asias-productivity-needs-a-boost-that-digitalization-can -provide#:~:text=Asia's%20Productivity%20Needs%20a%20Boost%20That%20 Digitalization%20Can%20Provide,-Amid%20slowing%20global&text=Asia's%20 strong%20eco.

Tourism Economics. 2021. "The Economic Impact of Online Travel Agencies in Europe 2019–2021." Oxford Economics, Oxford, U.K.

UNWTO (UN World Tourism Organization). 2023. "Tourism Statistics." https://www.e-unwto .org/toc/unwtotfb/current.

World Bank. 2021. "Antitrust and Digital Platforms: An Analysis of Global Patterns and Approaches by Competition Authorities." EFI Insight, Trade, Investment, and Competitiveness, World Bank, Washington, DC.

World Bank. n.d. *Mapping of Digital Along the Tourism Value Chain.* Unpublished.

WTTC (World Travel and Tourism Council). 2023. *Travel & Tourism Economic Impacts 2023: Global Trends.* London: WTTC.

Yang, Y., M. Nieto Garcia, G. Viglia, and J. Nicolau. 2021. "Competitors or Complements: A Meta-analysis of the Effect of Airbnb on Hotel Performance." *Journal of Travel Research* 61 (7): 1508–27.

Appendix G.
Country Restrictions to International Services Trade

Tables G.1 and G.2 report some of the most common restrictions to international services trade imposed by the larger countries in the East Asia and Pacific region.

TABLE G.1   International services trade restrictions, selected EAP countries

	China	Indonesia	Malaysia	Myanmar	Philippines	Thailand	Viet Nam
Fixed-line telecom	Com. Pres.; Other	Com. Pres.	Com. Pres.; Other	Other	Other	Closed	Com. Pres.; Res. Int.
Mobile telecom	Com. Pres.; Other	Com. Pres.	Com. Pres.; Other	Other	Other	Closed	Com. Pres.; Res. Int.; Other
Commercial banking	Com. Pres.; Other	Com. Pres.	Com. Pres.; Scope	Other	Com. Pres.	Closed	Com. Pres.
Life insurance	Com. Pres.	Com. Pres.; Other	Closed	Closed	Com. Pres.; Res. Int.; Other	Closed	Closed
Nonlife insurance	Com. Pres.	Com. Pres.; Other	Com. Pres.; Scope	Closed	Com. Pres.; Res. Int.; Other	Closed	Com. Pres.
Wholesale services	Other	Com. Pres.; Other	Closed	Scope			Com. Pres.; Other
Retail services	Other	Com. Pres.; Other	Closed		Other		Com. Pres.; Other
Maritime freight	Scope; Other	Scope; Res. Int.; Other	Com. Pres.; Scope; Other	Other	Scope	Closed	Other

Source: World Bank–World Trade Organization Services Trade Restrictiveness Index.
Note: "Closed" means that cross-border supply is not possible. Most of the time, the company must be incorporated/have its headquarters in the host country to supply services. "Com. Pres." means that establishment of a branch or a representative office is required to provide cross-border services. "Res. Int." refers to the requirement to use the services of a resident intermediary. "Other" includes sector-specific measures. "Scope" indicates limits on the scope of the service. EAP = East Asia and Pacific.

TABLE G.2   International services trade restrictions, selected EAP countries

	China	Indonesia	Malaysia	Philippines	Thailand	Viet Nam
Fixed-line telecom	FeIG (49); FeIA (49); JV; Other	Other	FeIG (70); FeIA (70); Screening; Other	Other	FeIG (49); FeIA (49); Man; Other	FeIG (49); FeIA (49); JV; Screening; Other
Mobile telecom	FeIG (49); FeIA (49); JV; Other	Other	FeIG (70); FeIA (70); Screening; Other	Other	FeIG (49); FeIA (49); Man; Other	FeIG (49); FeIA (49); JV; Screening; Other
Commercial banking	FeIA (25); Other	FeIG (99); FeIA (99); ENT; BoD; Other	ENT; Other	FeIG (Rec, 0); FeIA (Rec, 0); ENT; Other	FeIG (25); FeIA (25); Other	Screening; Other
Life Insurance	JV; ENT; Other	FeIG (80); FeIA (80); JV; BoD; Other	FeIG (70); FeIA (70); ENT	Other	FeIG (25); FeIA (25); JV; ENT	ENT; Screening; Other
Nonlife Insurance	ENT; Other	FeIG (80); FeIA (80); JV; BoD; Other	FeIG (70); FeIA (70); ENT	Other	FeIG (25); FeIA (25); JV; ENT	ENT; Screening; Other
Wholesale services		Other	Screening; Emp; Other		ENT	
Retail services		ENT; Scope; Other	FeIG (70); FeIA (70); ENT; Screening; Emp	Scope; Screening; Other	ENT	
Maritime Freight	Scope; Other	FeIG (OF, 49); JV (OF); Scope (OF); Other	FeIG (NF, 49); FeIA (NF, 49); BoD (NF); Emp (NF); Other		FeIG (NF, 75); FeIA (NF, 75); Scope (OF); Other	FeIG (49); FeIA (49); JV (NF); Man (NF); Emp (NF); Screening; Other

Source: World Bank–World Trade Organization Services Trade Restrictiveness Index.
Note: "BoD" indicates that a majority of the board of directors must be residents/nationals. "Emp" means that a minimum percentage of national employees is required. "ENT" means that the number of suppliers/licences is determined through an economic needs tests. "FeIA (%)" indicates a maximum aggregate foreign ownership allowed for the acquisition of an existing domestic entity. "FeIG (%)" indicates a maximum foreign ownership allowed in a new locally incorporated company. "JV" means that establishment as a joint-venture is required. "Man" means that managers must be nationals. "Other" includes sector-specific measures. "Scope" indicates limits on the scope of the service. "Screening" means that the investment is screened subject to evidence of economic benefits.

Appendix H. Quantifying the Potential Impact of Fully Implementing the Amendments to the Public Service Act

The elimination of foreign equity restrictions in the Philippines' Public Service Act (PSA) amendment can potentially cut foreign direct investment restrictiveness in the transportation and communication sectors by 75 percent. The reform would make these sectors some of the least restricted services in the economy. More open transportation and telecom sectors are likely to lead to improved services provisions, either through lower prices, improved quality, or increased varieties. These changes will benefit those sectors that most rely on transportation and telecom services as inputs.

Using the most recent input-output table for the Philippines (2018), we estimate the input share of liberalized services in downstream activities. We calculate for each sector of the Philippine economy the share of its input bill that is accounted for by transportation and telecom, and then identify how relevant foreign direct investment (FDI) restrictions in upstream sectors are for a given sector operating upstream. We match each input with its corresponding index score, and then calculate the weighted average score, using the share in the sector's total input bill as weight.

With these data, we perform a back-of-the-envelope calculation of the impact of a fully implemented PSA reform on productivity of downstream sectors. For these purposes, we rely on the elasticities of downstream productivity with respect to changes in FDI regulatory restrictiveness in upstream services sectors from Indonesia (Duggan et al. 2013). Because these elasticities are estimated by type of reform—including reforms that focus on changes in foreign equity limits—they are well suited to simulate the economywide impact of this reform if firms in the Philippines were to react similarly to Indonesian ones when they face changes in upstream FDI restrictiveness.

The PSA reform is estimated to increase total factor productivity by 3.2 percent on average, because it boosts competition in key enabling sectors and facilitates technology spillovers. The sectors with the largest shares of transportation and telecom in their input mix—metals and electronics—would have the largest productivity increases: should the PSA reform be fully implemented, the two sectors would experience double (6.4 percent) the average productivity boost of other sectors.

FIGURE H.1 Estimated impact of Public Service Act reform on productivity, the Philippines

Change in upstream FDI restrictiveness index*semi-elasticity of productivity to changes in FDI foreign equity restrictiveness (-0.975)

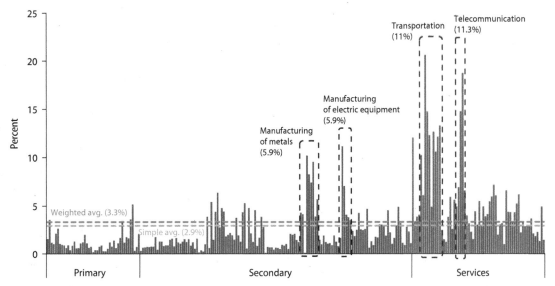

Sources: Based on 2018 Input-Output Accounts for the Philippines in million Philippine pesos; Organisation for Economic Co-operation and Development, FDI Regulatory Restrictiveness Index (https://www.oecd.org/investment/fdiindex.htm). https://psa.gov.ph/content/psa-releases-2018-input-output-tables.
Note: Warehouse, support activities for transportation, and postal and courier activities are assigned the average FDI index of transportation services. FDI = foreign direct investment; PSA = Public Service Act.

Reference

Duggan, V., S. Rahardja, and G. Varela. 2013. "Service Sector Reform and Manufacturing Productivity: Evidence from Indonesia." Policy Research Working Paper 6349, World Bank, Washington, DC.